Picturing China in the American Press

Lexington Studies in Political Communication

Series Editor: Robert E. Denton Jr., Virginia Tech

This series encourages focused work examining the role and function of communication in the realm of politics including campaigns and elections, media, and political institutions.

TITLES IN SERIES:

Governing Codes: Gender, Metaphor, and Political Identity, By Karrin Vasby Anderson and Kristina Horn Sheeler

Paving the Way for Madam President, By Nichola D. Gutgold

Maryland Politics and Political Communication, 1950-2005, By Theodore F. Sheckels

Images, Issues, and Attacks: Television Advertising by Incumbents and Challengers in Presidential Elections, By E. D. Dover

Democracy as Discussion: Civic Education and the American Forum Movement, By William Keith

Nuclear Legacies: Communication, Controversy, and the U.S. Nuclear Weapons Complex, Edited by Bryan C. Taylor, William J. Kinsella, Stephen P. Depoe, and Maribeth S. Metzler

Picturing China in the American Press: The Visual Portrayal of Sino-American Relations in Time *Magazine, 1949–1973,* By David D. Perlmutter

Picturing China in the American Press

The Visual Portrayal of Sino-American Relations in Time *Magazine, 1949–1973*

DAVID D. PERLMUTTER

LEXINGTON BOOKS
A division of
ROWMAN & LITTLEFIELD PUBLISHERS, INC.
Lanham • Boulder • New York • Toronto • Plymouth, UK

LEXINGTON BOOKS

A division of Rowman & Littlefield Publishers, Inc.
A wholly owned subsidiary of The Rowman & Littlefield Publishing Group, Inc.
4501 Forbes Boulevard, Suite 200
Lanham, MD 20706

Estover Road
Plymouth PL6 7PY
United Kingdom

British Library Cataloguing in Publication Information Available

Library of Congress Cataloging-in-Publication Data

Perlmutter, David D., 1962–
Picturing China in the American press : the visual portrayal of Sino-American relations in *Time* magazine, 1949–1973 / David D. Perlmutter.
p. cm. — (Lexington studies in political communication)
Includes bibliographical references and index.
ISBN-13: 978-0-7391-1819-1 (cloth : alk. paper)
ISBN-10: 0-7391-1819-6 (cloth : alk. paper)
ISBN-13: 978-0-7391-1820-7 (pbk. : alk. paper)
ISBN-10: 0-7391-1820-X (pbk. : alk. paper)
1. China—Press coverage—United States. 2. *Time.* I. Title.
PN4888.C54P47 2007
070.4'49320951—dc22 2006038447

Printed in the United States of America

∞™ The paper used in this publication meets the minimum requirements of American National Standard for Information Sciences—Permanence of Paper for Printed Library Materials, ANSI/NISO Z39.48-1992.

To two generations of Sinophiles:
my parents and my children.

Contents

List of Tables

List of Illustrations

Acknowledgments

Completion of this interdisciplinary project, which took some ten years, would not have been possible without the help of kind colleagues across several fields of study. Within mass communication I want to thank Tsan-Kuo Chang, Michael Griffin, Chin-Chuan Lee, John Maxwell Hamilton, Zoe Smith, Paul Lester, Jean Trumbo, Robert Denton, and Peggy and Melvin DeFleur. In history and area studies I want to register my appreciation for Scott Wong, whose encouragement helped renew my interest in writing this book; Ann Waltner, who started me in the direction of writing it in the first place; and John Henderson, who meticulously reviewed several drafts. A number of students and research assistants were very accommodating, especially in cultural and linguistic translation. First and foremost was my RA, Shannon Wang, but also helpful were Svetlana Kulikova, Amy Wang and Suping Zhang. In addition, I want to thank Joseph Parry (Lexington Books).

Furthermore, this book could not have been published without a grant from my former home institution—where I completed the bulk of the manuscript—the Manship School of Mass Communication at Louisiana State University. Additional funds were generously endowed by my present employer, the William Allen White School of Journalism and Mass Communications at the University of Kansas. Above all, I want to thank my proofreader and discussant-in-chief, my wife Christie.

Romanization of Chinese Names

Note: This history comprises a period where Romanized translations of Chinese proper nouns changed several times. So as not to confuse the reader, in almost all cases I have updated names using the contemporary "Pinyin" system. This decision may result in some historical instability—for example, Nixon and Kennedy debating the status of "Jinmen," not Quemoy—but I hope it will lead to overall clarity. The only exceptions are "Chiang Kai-shek," Madame Chiang, San Yat-Sen, Hu Shih, Chiang Ching-Kue.

Original	**Pinyin**
Chen Ye	Chen Yi
Chiang Ch'ing	Jiang Qing
Ch'ing Dynasty	Qing Dynasty
Chou EnLai	Zhou Enlai
Fujien	Fujian
Kiaochow Bay	Jiaozhou Bay
Kuo Hsi	Guo Xi
Kwangsi	Guangxi
Kwangtung	Guangdong
Li Piao	Lin Biao
Mao Tse-Tung	Mao Zedong
Matsu	Mazu
Nanking	Nanjing
Peking, Peiping, Pekin	Beijing
Peng Duhai	Peng Dehuai
Quemoy	Jinmen
Shantung	Shandong
Shenxi or Shensi Province	Shanxi Province
Sung Dynasty	Song Dynasty
Yenan	Yan'an

Prologue: Translating Images of China

> "You know the remarkable thing about China is that everyone who comes here becomes enchanted with the tremendous possibilities for achieving whatever it is they want to achieve."
>
> Henry Luce[1]

Journalism, it is often said, is the "first draft of history," but very few verbal descriptions of major news events survive the fading of newspaper pages. Pictures, however, can become mnemonic shorthands, whether as icons or in the stream of familiar "pictures of the past."[2] In *Picturing China*, my task is to assess photography and other images such as maps, charts, cartoons, paintings, and drawings of China within a chronological survey of the major events (and many minor happenings) over a crucial period of America's interaction with the People's Republic of China (and Taiwan). This sweep of 1949–1973 is conducted from the present, that is, using some of the latest sources and scholarship—the "what we know now" historical perspective. Then I evaluate, image by image, the view that the average American news reader would have had of America's China relations through one of the most important sources both of foreign affairs coverage and China news: *Time* magazine. Immediately, the present reader may offer that it is unfair to judge journalism of the past by the factual standard of half a century of scholarship and daylighted archives and memoirs. I agree, and I hope my own editorial stance here, which intrudes into the narrative occasionally, is one of sympathy with the reporter's task of sketching out the complexity of Sino-American relations based on extremely limited knowledge.

Yet one key observation makes this assessment also partly an indictment: at almost no point in the decades of news coverage of China did the editors of *Time* give more than cursory notification to their readers that the facts and images and opinions they offered—the proverbial "first draft of history"—might have been incorrect, misleading, or fantastic. My chronological approach proved to be illuminating in revealing the curious incongruity that *Time* would frequently provide its "expert" testimony about what was going on in China only to be proven wrong later, but never admitted its own fallibility. That is not to say that this book is an attack on *Time* magazine, journalism, or foreign affairs reporters or editors. It is a critique in the true sense of the word. I hope to present some insights and data that I think help us better understand the tremendous difficulty and many subtle nuances of creating accurate, reliable, and meaningful images of other countries, peoples, and far-off events.

My justification of the venue is that *Time* was and is both the biggest-selling American newsmagazine and arguably the best determinant of what is in the news and which pictures of the events are in the forefront of the journalistic agenda. Its images are good indicators of the pictures in circulation about China from year to year, but they are also representative of a publication whose owner, Henry R. Luce (who died in 1967), was intensely concerned about China and who had strong opinions on what kind of China he wanted readers to know about. As I found in an earlier study, *no American news platform had more print word and images about China than* Time.[3] It was also Luce who, in a famous essay in 1941, proclaimed that the twentieth century was "America's first century as a dominant power in the world."[4] The "Lucepress," as it came to be known (the publisher also controlled *Life* and *Fortune*), worked relentlessly in the late 1930s and through 1941 to build American sympathy for China as a victim of Japanese aggression. Such sentiment spurred the United States to enact embargoes on sales to Japan of critical resources such as oil, which in turn set the stage for the attack on Pearl Harbor. In addition, it was Luce and other friends of China like Pearl S. Buck who would push for the repeal of the Exclusion Act in 1943, which had for more than half a century banned Chinese immigration to the United States.[5]

In the late 1940s and '50s, the Lucepress led the charge in accusations against State Department officials of being pro-Communist and of "losing China."[6] Luce also was the "visible headman" in the early days of the China Lobby, which sought until the 1970s to stop America's diplomatic recognition of Communist China.[7] At the same time, we must credit Luce with not allowing antagonism toward China in the 1950s to translate into racism against Chinese Americans at home. And he did have the attention of the high and the mighty. President Kennedy held a subscription to *Time*, believing that it "reached more influential members of the nation living outside of Washington and New York than any other news service."[8] The ideological implications of Luce's Nationalist bias in Taiwan—he once pledged that his publishing empire would attempt to defeat international Communism—are considerable.[9]

As an editor, Luce was complicated. He tended not to give direct comments to his staff members to point a story in one direction or another; rather, he often asked their opinions, dropped hints, or made strong arguments with which he hoped they would ultimately end up agreeing. As one longtime friend and editor once chided him, "Harry, I wish you wouldn't give us as much argument. Why don't you just give us a few orders?"[10] Luce also hated sycophancy and tended to be so suspicious of yes-men that Hedley Donovan, who took over the reins of *Time* when Luce stepped down in 1964, complained that his boss loved "argument so ardently that it takes a certain amount of intellectual courage to agree with him when he is right, as is bound to happen from time to time."[11] For example, *Time*'s leading correspondent in China for many years, Teddy White, was the ideological opposite of Luce. The Boss was certainly not a one-dimensional ideologue. The year before his death, the publisher petitioned to visit the PRC.[12] Furthermore, an analysis of *Time*'s editorials found admissions of problems among the Nationalists and potential for the Communists.[13] Indeed, my study shows that while there was a strong ideological bent (and even cant) in images of China in *Time*, most noticeable in the Luce era of supervision were complexities and contradictions.

Time, thus, is: (1) an intrinsically interesting case study, fascinating for its particular status as a representation of the specific influences coming together under Luce's direction; (2) a leading American newsmagazine, perhaps not representative of coverage in other publications but of special importance nonetheless due to its influence as an "opinion leading" and "agenda setting" publication; (3) representative of U.S. news coverage more generally, a reflective index of American images of China as a gauge of America's perception/imagination of China.

Although I approach the American view of China chronologically, certain themes emerge. One is China's "mass," often distinguished not just as a big group of people or a large swath of land but as further symbolism of some quality about China. The people of China could be perceived as an opportunity: missionaries might see them as a mass of souls who required saving; business interests would look upon the Chinese multitudes as a trading, investing, and labor opportunity; American laborers might see them as fatal competition. China's "mass" could also refer not to her people but to her resources, her wealth; when Marco Polo raved about China's "millions," he meant the number of gold pieces in her treasury. The Chinese mass of population could be seen as a threat, as in the Yellow Peril vision of the turn of the century or the hordes of the Chinese "volunteers" in Korea, human wave battalions attacking American soldiers with a great stamp of feet and a bugle call. Or they could be the perceived threat of the Chinese mass army that so highly influenced the course of America's Vietnam War strategy. Soviet premier Nikita Khrushchev, angry at Mao Zedong and his regime in a period of Sino-Soviet hostility, would describe China as "a mass of human flesh and nothing else." But Richard Nixon, in 1967, arguing that America should reach out to China, would point to the mass as a reason for rapprochement: "There is no place on this small planet for a billion of its potentially most able people to live in angry isolation." At last, when the president did travel to China, a *Time* cartoon transmographied the Chinese mass into a pop culture incarnation of '70s Americana, where Nixon and Mao shake hands amid a closely packed assembly of both slant-eyed and round-eyed smiley faces.

Obviously, such an overview of China images for some twenty-five years can be organized in several different ways: by themes, for example, or clustered around great events. But I chose to create dual chronologies. Each chapter contains two major sections. First, I offer the "what we know now" examination of history, a narrative and analytical description of the events of certain eras of Sino-American relations, such as, for instance, the Taiwan Straits crises of the 1950s. My goal here is to explain in detail to the reader what, as best as we can tell from present sources, was "really" going on. I admit candidly that scholarship marches on, and so do the prejudices of the contemporary, so that a future historian might offer a different version of, say, the Korean War than that which I have provided here. The second part of the picture-study chapters is the chronological description and analysis of the images—the maps, the photographs, and so on—of and about China. As I often will note, many of these images have discordances with the narrative history to which present sources testify. And that is a key point: I want to illustrate how *Time*'s editors showed China, from 1949 to 1973, and how *Time*'s readers would have seen, week by week, the panorama of the Middle Kingdom.

The result: a portrait of the changes in the way China was pictured via *every picture* in a publication that represents one important slice of the data universe that almost all Americans who were not area experts or political leaders probably saw as portraying China.

The questions I wish to answer about these pictures are:

- What did we see?
- What did we not see?
- What were we told the pictures meant, symbolized, or stood for?

In looking at an image in its context, historical background, cultural allusions, and words surrounding it on the page will be important for what I hope to accomplish here. In our image-dense society, we rarely encounter any picture in isolation from others, whether on a magazine page or in the flitting of a television screen. Pictures conceptualize and contextualize other pictures; so do words on the sound track or the voice-over of the documentary or the caption of the news photograph. Typically, in news, the purpose of pictures is to illustrate words: an item on flooding in China's Hubei Province of is accompanied by a picture of storm waters engulfing a quay with sampans, for instance. A few images, then, serve as metonyms for the greater story. And captions and context affect what we think about an image's meaning. For example, *Time* of January 5, 1970, contained the following ironic item: "Vatican officials were taken with an anonymously-donated portrait of a handsome young man in a windblown, cassock-like robe. In their eyes, it symbolized a devoted missionary priest, and they hung the painting near Pope Paul's own likeness in the pressroom of the Holy See." Suspicions about the painting's origins mounted, however, and later the Vatican removed the picture when "a little investigation revealed [the artist] had copied a portrait of Chairman Mao as a youth of twenty-seven, striding through Jiangxi Province in China's traditional gentleman's robe."

For decades of covering China, however, publications like *Time*, having no reporters in the country, could only reprint material from PRC government media such as the Eastfoto news service. But *Time* was anti-Mao and anti-Communist; simply posting "Red" propaganda with neutral captions was ideologically impossible. Instead, *Time* used its captions to counter the images it described—that is, the visual content is mocked, questioned, discounted by the surrounding words. So an intended-to-be-flattering PRC-created image of Mao might be labeled "pudgy dictator." Captions also serve to certify what we are seeing in more subtle ways. *Time*'s argument throughout much of the early part of the Korean War was that China was a puppet of Moscow and would not intervene to help North Korea. To visualize this "fact," maps of North Korea's borderland show "Manchuria" (often crossed by "Soviet" supply lines), but nothing labeled China, although at the time Manchuria was China. Out of cartographic sight was a reflection of the ideological mind; hence the shock when hundreds of thousands of Chinese troops crashed into American forces on the Korean Peninsula. Finally, the wider frame through which we see an image sets off, or primes, many preexisting associations that help us construct meaning. In 1954, a picture of a "dead nationalist soldier killed by Communist shelling" would have triggered a wholly different reaction from the American public than a picture of the same body of an Asian man labeled "Dead Communist Soldier killed by Nationalist shelling." Meaning—as well as sympathy and empathy—is in the mind of the beholder of all images. In sum, pictures matter, but so does context.

Also, the personal perspective of these public images is difficult to overestimate. Writing at the time of the Tiananmen events, journalist Jay Mathews noted that "the emotional commitment to China remains among today's correspondents, particularly for those of us who fell in love with the romance and intrigue of China's history, culture, and size when we were still in school."[14] Such a statement could have been written in the 1930s and 1940s as well. A noteworthy phenomenon about images of China that permeates this analysis is one suggested many years ago by the late political scientist Alvin Z. Rubinstein. He commented that, in his experience, "people in the West who studied China tended to love the Chinese people, while people who studied Soviet Russia tended to be indifferent or actually dislike the Russian people." He did not mean this observation to be taken as scientific fact, but in general a preponderance of Westerners who have influenced our visions of China—from Marco Polo to the Jesuit Fathers to Pearl S. Buck, Henry Luce, many of the "China Hand" diplomats and reporters of the 1930s and 1940s, and most students of China in the academy today—loved or love China. They may have found fault in political situations in China, governments of China, or even aspects of the Chinese character, but basically they were or are fascinated and romantically attached to the Chinese people and culture. Many Chinese leaders, through the present day, have found such "*lao peng you*" (old friends) useful for affecting U.S. policy toward and press portrayal of China.

Further, I would be misrepresenting myself if I pretended to be a dispassionate, objective observer of this phenomenon and not a co-conspirator. A Sinophilic academic and son of two Sinophilic academics, I have been interested as far back as I can remember in everything to do with China and the Chinese. Indeed, endless dinner-table discussions, a Chinoiserie-permeated house, and a professional career that included "adopting" half a dozen Chinese graduate students has planted in the next generation of my family an ingrained Sinophilia. How this predilection affects my own analysis of other people's portrayals of China in the ninteenth and twentieth centuries I leave to the reader to decide, although at times I will attempt self-referential comment. Suffice it to say that I agree with the scholar Tu Weiming that nonethnic Han China scholars and journalists are part of the "cultural China" that they study.[15]

One other controversial decision I have made for this book is that it is largely theory-less. Originally, there were several chapters of discussions of theories of public opinion and the press, agenda setting, framing, and speculations on the relationship between foreign policy and foreign press coverage. Because of the size of the study itself, I have excised these portions and refer the reader to other writings on these theories. Basically, as is developed here, in many cases the press coverage in *Time*—that is, the image of China—mirrored American foreign policy toward China, but many other times—such as, for example, during the revolution and the latter part of the Korean War—government policy was thoroughly pilloried. I make my own speculations and observations, but I will note when *Time* was watchdog, lapdog, or attack dog. The theory, I believe, is revealed within the study and this research is little served by overwhelming the reader initially because the whole point of my exercise here is the necessity of immersing oneself in the data about China to get anywhere closer to understanding not only "what was going on in China" but "what we saw about China."

Finally, there is a very curious blowback to our coverage of China, which I occasionally explore within these pages. There have been a number of excellent dissertations and a few monographs on the subject of Mao, the most important man and policymaker in China during the entire scope of this study, and his enduring fascination with America. It is not accurate to trot out the old cliché that Mao had a love-hate relationship with America. Certainly, Mao at different points of his life was puzzled by, infuriated with, or admiring of the United States of America; there is no evidence that he ever hated America in either a racial sense or a national one. Indeed, as was said above and as will be repeated below, the policy line of *Time* magazine was that any antagonism or animosity was between national leaders, not between peoples. Someday, however, some Chinese scholar will write an excellent book on Mao's relationship with the American press. We know from inner-circle sources like Mao's personal physician that the former paid close and keen attention to American newspapers and magazines (in translation), including, presumably, Luce's *Time*. We know that in several cases, such as in the tricky diplomatic wrangling after the fall of Nationalist China, Western news reports were taken by Mao to be representations of White House foreign policy. (Many U.S. administrations recognized this problem. When Nixon went to China in 1972, one of his most intriguing statements to the Chinese leaders was, in essence, "Don't pay attention to the press. They don't speak for us.") So although this book does not focus on the subject of whether *Time* magazine and the American press in some ways set the agenda for Chinese foreign policy, the kernels of such a study are presented in the data here.

Let me give one introductory example to the reader. In the chapter on the Korean War and the mutual debacle of America's undeclared war with China in the peninsula, there is extensive discussion about America's reading, or misreading, of Chinese intentions about whether to intervene in Korea, as well as the Chinese problematic reading of American foreign policy objectives and parameters. We know now from a variety of Chinese sources (although many more still remain classified) that the Chinese strongly believed that American intervention in a civil war between North and South Korea was simply a pretext and a prelude to a direct invasion of China itself. It seemed logical from the Chinese perspective, then, to launch its armies into the peninsula and thus establish Korea, and not Manchuria, as the decisive battleground. In short, it is clear that Mao and many of his comrades held an early-1950s version of the domino theory, with China as a target of aggressive American imperialism. In an August 4, 1950, Politburo meeting, Mao said, "If American imperialism wins, it will be proud and threaten us. Thus we must help North Korea."[16] His chief diplomatic minister Zhou Enlai echoed in a National Defense conference of August 26 that "The American imperialism intended to open a crevice at Korea as the base for a new world war. . . . If [the] U.S. oppressed Korea, undoubtedly, its next step is to oppress Vietnam and other former colonized countries. Therefore, the Korean War is the focus of the occidental struggle at least."[17] Later in the year, Zhou reasserted that "Chinese people can't stand for foreign invasion; nor will they let alone imperialists to invade the neighbors."[18] After the China-America combat had begun, Zhou affirmed that the "Korea affair is just an excuse; even if not in Korea, the war would break out at other places, such as Taiwan, Vietnam . . . For us, the Korean war is the most beneficial."[19] And every time that Mao spoke about the "whole occidental situation," he put it in terms of an American dagger pointed at the heart of China—that is, Manchuria.[20] I argue below that such paranoia was well founded; if Mao had been reading *Time* magazine, he would have been under the paradoxical impression that (a) the United States dismissed China as a nonentity in Korean affairs, and (b) China posed a terrible threat to its neighbors.

Picturing China, then, is an attempt to assess China pictures within an empirical framework, not for specialists only but for students and the interested public affairs reader. I hope this book will make a contribution to the public debate on the question of how the press does or should influence the making of foreign policy. In an age when the camera is the default (and much faulted) medium of international relations, such a study can provide a useful perspective, even in retrospect.

In this light and to this end I have organized the book as follows:

Chapter 1: Opening China: War and Revolution, contextualizes and begins the main analysis of the book. The year 1949 was a significant one for America-China relations: the Nationalist government collapsed and fled to Taiwan; Mao Zedong established the People's Republic of China (PRC); U.S. relations with the new Communist government spiraled from lukewarm to hostile, setting a pattern for decades to come; and a new China image began to emerge—China as menace, as threat to world order and American democracy. In the rubble of the "fall of China" debate at home, China also became a domestic issue, with conservatives accusing the Truman administration of "losing" China. Many historians see this struggle as the first drumbeat of McCarthyism. At the same time, China became an ally of the Soviet Union, although the relationship was not as straightforward as Cold Warriors of the West assumed. Meanwhile, pictures of "China" became those of the "two Chinas"—the PRC and Taiwan—and fewer journalists found access to the Communist mainland. It was during this time as well that the long-term "recognition" issue gestated: Who comprised the legitimate (and U.N.-seated) government of China? The chapter explores what Americans were shown of these tumultuous events, and how these images were captioned and framed for the viewers' consumption.

Chapter 2: The Korean War, 1950–1953, examines the visualization of the first great armed clash between China and the United States. From the prequel of the U.S. government publicly stating that Korea was not in its sphere of influence to the U.S.-led coalition to stop North Korea's invasion of the South to the fateful decision to drive north toward the Chinese border to the Chinese counter invasion and long bitter war of attrition, we find a clear case of the limits of pictures. The major story of the Korean War was how American leaders and public were twice surprised (by the initial attack by the North on the South and by the Chinese intervention) despite innumerable warning signs and portents, some of them quite blatant. American journalists followed the guidance and wisdom of the government, and the result was an "intelligence failure" that was undergirded by a failure of photojournalism. The cost was 53,000 dead Americans and deepening bitterness in Sino-American relations.

Chapter 3: Armageddon for Jinmen?: The Taiwan Straits Crises, 1954–1959, focuses on the genesis of America's "special" relationship with Taiwan. I examine how the island was transformed in visual journalism from insignificance to being the centerpiece and bastion of U.S. Far-East polices and plans, to the point where twice in the 1950s the Eisenhower administration seriously considered "going nuclear" in its defense. Taiwan, as we see, was used as a "good" China contrasted to the "bad" Communist mainland, a visualization that entailed actively covering up oppressive aspects of the Nationalist regime on the island. Here I assess one of the problematic features of photojournalism—not allowing us to "see through" the eyes of our enemies in gauging our own actions.

Chapter 4 covers the period of interlude and stasis, 1960–1967. The 1960s saw no open conflicts between the United States and China, yet the "Bamboo Curtain" between the two nations seemed unyielding—especially so when China was increasingly seen as the supporter of the North Vietnamese and Vietcong. Very little personal contact occurred between Chinese and Americans, and China became more inward looking while in the throes of the Cultural Revolution. Yet, in the Kennedy and Johnson administrations, a pattern was established regarding China policy: the incoming president would foresee a new way of dealing with China, then domestic politics and international incidents would erupt to deflect and stifle such hopes.

Chapter 5: Journey to China, 1968–1973, examines the period when the American internal consensus to isolate China gradually declined, and incoming president Richard Nixon, his anti-Communist credentials impeccable, saw China as a possible quasially in the Cold War. As Americans were becoming aware, there was an increasingly violent Sino-Soviet split. This chapter looks at the long march to Nixon's visit in 1972, and especially at the trip itself and what it revealed about the way we looked at China through new hopes and fascinations. Of special interest is the "reversal" of China's image during Nixon's visit, from vilification to adoration.

Finally, I sum up the book's findings, arguing that *Time* magazine systematically attempted to skew, recode, redirect, or overturn for persuasive ends the average American's image of China. Pictures, thus, were not just passive illustrations, but framed how editors believed we *ought* to have perceived China. Despite the slogans of modern newsmakers who promise to show us "the world," this study provides evidence that the way we saw China in the late twentieth century (and how we see it in the twenty-first) is no less complex or more straightforward than it was in the past.

Notes

1. Henry R. Luce, "Four Messages from China" (June 1941 speech for United China Relief), in *The Ideas of Henry Luce*, ed. John K. Jessup (New York: Atheneum, 1969), 196–97.
2. David D. Perlmutter, *Photojournalism and Foreign Policy: Framing Icons of Outrage in International Crises* (Westport, Conn.: Greenwood, 1998).
3. David D. Perlmutter, "Visual Images and Foreign Policy: Picturing China in the American Press, 1949–1989" (Ph.D. diss., Minneapolis: University of Minnesota Press, 1996).
4. Henry R. Luce, "The American Century," *Life*, February 17, 1941.
5. Xiaohua Ma, "The Sino-American Alliance During World War II and the Lifting of the Chinese Exclusion Acts," *American Studies International* 38, no. 2 (2000): 39–61.
6. W. A. Swanberg, *Luce and His Empire* (New York: Charles Scribner's Sons, 1972), 314.
7. Swanberg, *Luce and His Empire*, 452–53.
8. J. L. Baughman, *Henry R. Luce and the Rise of the American News Media* (Boston: Twayne, 1987), 185.
9. Swanberg, *Luce and His Empire*, 414–15.
10. Thomas Griffith, *Harry and Teddy: The Turbulent Friendship of Press Lord Henry R. Luce and His Favorite Reporter, Theodore H. White* (New York: Random House, 1995), 90.
11. Griffith, *Harry and Teddy*, 92.
12. P. Neils, *China Images in the Life and Times of Henry Luce* (Savage, Md.: Rowman & Littlefield, 1990), 286.
13. Neils, *China Images*, 240–49.
14. Jay Mathews, "Edgar Snow Told You So; Why China's Great Leap Backward Should Come as No Surprise," *Washington Monthly* 21, no. 6 (July–August 1989): 50+.
15. W. Tu, "Cultural China: The Periphery as the Center," *Dædalus* 120 (1991): 1–32; Paul A. Cohen, "Cultural China: Some Definitional Issues," *Philosophy East & West* 43 (1993): 557; Chin-Chuan Lee, "Ambiguities and Contradictions: Issues in China's Changing Political Communication," in *China's Media, Media's China*, ed. Chin-Chuan Lee (Boulder, Colo: Westview Press, 1994), 3–22.

16. Bo Yibo, *Some Decision Making in Retrospect* trans. Suping Zhang. Zhong Gong Zhong Yang Dang Xiao Press, vol. 1 (Beijing: The Party School of the CPC Central Committee Press, 1991), 43.

17. Zhou Enlai, "Be Prepared to Win," trans. Suping Zhang, in *Selection of Zhou Enlai's Essays*, vol. 4 (Beijing: Renmin Press, 1997), 43–44.

18. Zhou Enlai, "Struggle To Solidify and Develop People's Victory," September 30, 1950, trans. Suping Zhang, in *Selection of Zhou Enlai's Essays*, 2:37.

19. Zhou Enlai, "Assisting Korea to Fight U.S. Is Still the Primitive Mission in 1951," April 2, 1951, trans. Supping Zhang, in *Selection of Zhou Enlai's Military Essays* (Beijing: Renmin Press, 1997), 4:181–82. See also: "The Claim of Zhou Enlai on the U.S. Military Invasion to Taiwan," June 28, 1950, in *The Volume of Files on Sino-US Relations* (Beijing: World Knowledge Press, 1960), 2:91–92.

20. Mao Zedong, "The Strategies of CPVA," October 2, 1950 trans. Suping Zhang, in Selected Military Writings of Mao Zedong (Beijing: The PLA General Staff Department, 1961), 6:106–8.

CHAPTER 1

Opening China: War and Revolution

> "When China is opened to Europeans, they will find it the finest model of administrative centralization in the world."
>
> Alexis de Tocqueville[1]

1. Introduction: Cycles or Themes?

China has a long history of relations with the West, from Roman silk traders to Marco Polo to court Jesuits, but the United States of America's encounter with the so-called Middle Kingdom began before the former was even an independent nation. In his history of Sino-American interaction during the European Enlightenment, Aldridge observes, "During colonial times, China existed in America as a mirage or a reflection of the attitudes and images of European writers, the majority of whom had never seen Asia themselves."[2] What they saw or heard of was a fabulously large, wealthy, and peaceful kingdom; to this, Europeans contrasted the squabbling and warring monarchies of their own continent, while Americans imagined the corresponding possibilities of their young republic. By the nineteenth century, China presented a much less appealing image to the West. The Chinese regime was collapsing at the same time as Western expectations altered. As historian Jack Gray points out, "The decisive factor in Western views of China was not change in China, but change in Europe."[3] Before, China seemed a benevolent kingdom, a silken leviathan run by enlightened, educated, cultured elites; but now, "China had come to symbolize almost the opposite: a polity stifled by the power of a self-perpetuating elite who were both guardians of a quasi-religious tradition and the servants of an arbitrary despot."[4] Among the complaints about China were that it did not open itself up to full trade with the West; it suffered through civil war and seemed helpless against foreign enemies; it refused to cast off its archaic traditions; it balked at rapid modernization (i.e., to become Western); and, most of all, it did not embrace Christianity.

China, thus, was in many ways a ready foil for American self-inspection as well as expansion. Americans saw themselves as a new, young, vigorous power, the leader of a new world. China served well as an example of the decrepit vestiges of hidebound tradition. In the language of visual communication, the associative juxtaposition of "old, weak, dirty, corrupt" China was a perfect polar opposite for a healthy, strong, clean America. Americans used China to define themselves. As a student of late nineteenth-century American diplomacy in China comments,

> Most Americans had never seen China . . . but they believed that it lagged far behind the United States. Because American citizens conceived of their own nation as the epitome of progress and freedom, they assumed that China with its archaic imperial government, and its hoary Confucian traditions, was the antithesis of all the United States had achieved.[5]

Americans, proud of their egalitarian and democratic heritage, liked to think of themselves as patrons of the oppressed, dissimilar in this respect from the European powers. However, in practice, American trade often relied on concessions won by Europeans in China, especially the English.[6] Instead of directly confronting or warring on Chinese authority, Americans demanded a share in any treaty rights wrung from the dying empire.

In the twentieth century, American attitudes toward China were governed not just by a craving for self-validation, but also by the actions of other countries. When China was seen as a victim, such as in the conflicts with Japan in the 1930s, American opinion, always registering a tendency to favor the proverbial underdog, was sympathetic toward China.[7] But we must be judicious when speaking of "eras" of American views of China. The eminent French historian Marc Bloch once warned against the follies of "hectohistory"—that is, artificially dividing up history into periods that happened to coincide with our dating system's predilection for hundreds, so that, for example, we speak of the seventeenth century as being different in some way from the eighteenth century. But even contending that any one block of time is a "period" is problematic, because the shifts in our "visions" of other countries are rarely rapid and often conflicting.

Most work on "images of China" follows Harold Isaacs, from his excellent book on American attitudes toward China, in defining several eras of the subject[8] (see table 1.1).

Table 1.1. Extension of Isaacs' Chronology of Western Attitudes toward China

Period	*Type of Attitude*	*Description of Attitude*
1784–1840	*Deference*	Appreciation of the grandeur and wisdom of Dynastic China
1841–1900	*Contempt*	Exasperation at China's failure to modernize, perceived corrupt autocracy, moral failings, but still an undercurrent of hope that a great awakening was imminent
1901–1950*	*Paternalism*	Missionary crusade to remake China in our (the West's) own image
1951–1971	*Fear*	Communist China as a danger, a threat to American safety, democracy, and capitalism
1972–	*Respect*	Respect for Communist China's increased strength and its power in world affairs, growing capitalist orientation and trade

*William Huntzicker (University of Minnesota) has suggested that this chart fails to take into account missionary activity that began in the 1930s. 1995. Private conversation.

Such a typology is useful, but only if we recognize its limits. For example, Americans did not suddenly sneer at China in 1842 while in 1839 they had admired it.

Even at the time when the first American trading vessel sailed to China (1784), the nation was by no means completely an unknown or universally beloved land among people living in North America. The editor of *Boston* magazine of the period, looking back at fifty years of travel reports about China, noted:

> Some travelers have extolled them [the Chinese] in the highest terms, whilst others have run into the contrary extreme; those last have been of late more numerous than the former. When we hear the character of a private person loaded with reproach, founded on real or pretended facts, we have no means of contradicting, though unknown to us, it is natural, for the sake of human nature, to wish that they may be true. But if we happen to be furnished with indisputable evidence of the virtue of that person in other instances, a mind of any benevolence will seize with pleasure and an opportunity of urging them in his vindication: much more should we be influenced by this principle of equity, when a large and populous nation is concerned?[9]

A sound point! China was (and is) big enough for almost anything to be said about it. So the same Chinese character types, holding the same social positions, espousing the same values, performing the same actions, could be held in inverse attitudes. Indeed, it is more accurate to say that China held certain "themes," all of which actualized coexisting "images," some more powerful than others at certain times. Isaacs understood this, writing, "The Chinese are seen as a superior people and an inferior people; devilishly exasperating heathens and wonderfully attractive humanists; wise sages and sadistic executioners; thrifty and honorable men and sly and devious villains; comic opera soldiers and dangerous fighters."[10] Americans could simultaneously sympathize with a China preyed upon by Japanese militarists and practice the worst sort of prejudice against a Chinese woman living in an American city.

Although both negative and positive attitudes toward China coexisted even in ages of "respect," the negatives tended to outnumber the positives in the middle and late nineteenth century. Current events—such as the failed diplomatic missions to China, the Opium Wars, and the decay of the Qing Dynasty—had an influence on views both good and bad. In addition, many Westerners were traveling to China for trade or as missionaries, and their reports varied. At home, Americans would have read more than ever before about China in the newspapers and been exposed to China in popular culture and museums, and at traveling shows. Finally, the Chinese were coming to America themselves in greater numbers beginning in the 1850s. Then, a series of events, including the Republican Revolution of 1911, the rise of warlordism, the attacks on China by Japan, and finally China's becoming a U.S. ally in World War II combined to support China visions that were increasingly sympathetic toward China, if not her leaders then at least for her suffering people. A key factor, especially in the 1930s and beyond, was that the American government and powerful media owners such as Henry Luce—publisher of *Time*, *Life*, and *Fortune*, and son of Presbyterian missionaries in China—actively worked to bolster China's image. Chinese leaders, like the American-educated wife of the president of China, would be crucial (and visually distinct) "China shapers" for the American mind. During World War II, there would be an even more organized, sustained, and effective "pro-China" propaganda effort aimed at Americans.

2. Propagandizing China: The War Years

Premodern American views of China were formed via many routes and sources. American missionaries in China—like Luce's father—were active proponents of "saving" the Middle Kingdom from many terrible fates, ranging from heathen folly to European exploitation. From the days of Marco Polo, China had the reputation of being a place of "massness"—huge populations, substantial wealth, great opportunity. During the 1930s, public opinion polls showed another important development: sympathy. China was seen as an underdog—which indeed it was—attacked by an aggressive, warlike Japan. The sharpest delineation began on September 18, 1931, when, some would argue, the first shots of World War II were fired; this was the "Mukden Incident."[11] The Japanese local "Kwantung army" accused the Chinese army of placing bombs that damaged the railway tracks of a Japanese-owned railway on the outskirts of the city of Mukden in Manchuria. Within hours, Japanese troops attacked the city and by March 9, 1932, had conquered all of Manchuria. Japan quickly declared the territory a dependent state, Manchukuo. Two years later Japan established the deposed Manchu emperor Pu Yi as its ruler. The entire incident, however, had been plotted by factions within the Kwantung army as well as others in Japan, but without the approval of many commanders in Tokyo or of the government. The confusion showed that Japan's government was not unified: the civilian leadership continued to oppose military aggressions; ministers were regularly assassinated; the army and navy were at odds. In the end, however, the ultranationalists who supported conquest won out. Of significance for China was that Chiang Kai-shek offered minimal resistance, believing that China was not prepared for a foreign war while he was still fighting a civil war with the Communists.

Mukden was also a test by Japan of the power of the League of Nations and America: Would the world enforce its principles to "end war" by means other than moral suasion? On both counts aggression seemed to be rewarded, since Japan held on to Manchuria until the end of World War II. A year after the region's conquest the League did condemn Japan's actions, yet the United States and Great Britain decided not even to employ economic sanctions. But the Manchuria crisis did begin to establish that Japan was the "villain" in the relationship, with China as the "good guy" in the minds of most of the American public.

The undeclared war, which the Japanese euphemistically referred to as the "China incident," convinced President Franklin Roosevelt that drastic measures must be taken, and so the long chain of economic sanctions against Japan and aid to China began. America was the leading supplier of munitions to Japan, including many items like high-grade steel that were essential to modern combat and could be obtained only from the United States.[12] At the same time economic assistance was stepped up to China, at least indirectly, through, for example, the purchase of Chinese silver. But Japanese provocations such as the Panay incident continued. After an extensive series of battles, Japan captured Nanjing (the "new" capital of China); the ensuing orgy of destruction came to be known as the "Rape of Nanjing." Atrocity images were common. The verb "rape" would continue to sum up the portrayal of what Japan was doing to China, and in the era of World War II it would be literally acted out in many fictional and documentary films. Then Japan set up a puppet Chinese government in Nanjing in March 1940; it was clear that the Japanese aimed to conquer all of China. Days later, the United States began direct aid to the Chinese government in Chongqing worth $20 million.[13] A series of progressively larger loans was extended, reaching $100 million in November.[14] American support for China and embargoing of Japan was a direct cause of Japan's decision to strike at American forces by the winter of 1941.

Unsurprisingly, the Japanese attack on Pearl Harbor was seen by many Chinese and friends of China in America as cause for a new and better chapter in Sino-American relations. For American missionaries, who for years had advocated help for the beleaguered Chinese, it was almost a sign of divine intervention that their original homeland and the people they had dedicated their lives to helping were now allies.[15] In China, the reaction to the news was relief that America was now "in the fight."[16] Han Suyin, the Chinese novelist (and no friend of the Nationalist regime) described the scene in Chongqing, the inland capital city of Chiang Kai-shek's regime, as follows:

> The Kuomintang government officials went around congratulating each other as if a great victory had been won. From their standpoint, it was a great victory, what they had waited for, America was at war with Japan. At last, at last, America was at war with Japan! Now China's strategic importance would grow even more, American money and equipment would flow in; half a billion dollars, one billion dollars. . . . Now Lend-Lease would increase.[17]

The expectation was that America, the arsenal and treasure chest of democracy, would flood China with guns and capital. For the Nationalist leadership—and not a few Communists—the internal struggle was now the most important battle because it was assumed that Japan was doomed. Chiang Kai-shek firmly believed in this outcome, and all through the war years he would frustrate a succession of American military and diplomatic representatives who could not understand why he seemed to favor fighting his Chinese enemies instead of Japanese occupiers. A further consideration supported Chiang in such a strategy. China had been absorbing the might of Japan for many years, with little outside help. That now the United States and Britain must take up the real burden of the war seemed only logical.[18] The day after the attack on Pearl Harbor, the Chinese held two meetings with British and American representatives in Chongqing, during which Chiang proposed a grand alliance focusing on the throwing back of the Japanese from China but also joint hegemony over the entire Pacific.[19] The implication was that of the three arenas of war the Allies would have to choose among to focus their efforts—Europe, the Pacific Islands, and the East Asian mainland—the last was the most accessible.

Many Americans shared the hope for the United States to invest heavily in China's war and future. Certainly, friends of China like Henry Luce were buoyed by the alliance. The publisher's father died the day after the Pearl Harbor assault, and when *Time's* China correspondent, Theodore White, gave his condolences to his boss, the younger Luce countered that he was glad his father "lived long enough to know that now China and America are both on the same side."[20] The publisher and other supporters of China like novelist Pearl S. Buck would advocate various China causes and aid projects, including the repeal of the Exclusion Act in 1943.[21] This feeling was shared, in theory and sentiment, by many others. Once again the appeal to the utility of China's "mass" was enticing. On December 9, the *New York Times* made the case that if the United States cooperated with China, "a loyal ally with . . . inexhaustible manpower," it would have "the key to the strategy of the Pacific."[22]

President Roosevelt and many in his administration were not immune to such musings—to the consternation of certain other world leaders. In an early meeting with Roosevelt, Winston Churchill remarked on what he felt was an inordinate fondness for China and commensurate magnification of its importance in the U.S. capital: "I told the President how much I felt American opinion overestimated the contribution which China could make to the general war. . . . I said I would of course always be helpful and polite to the Chinese, whom I admired and liked as a race (and pitied for their endless misgovernment) but that he must not expect me to adopt what I felt was a wholly unreal standard of values."[23] Later Churchill stated, "At Washington I had found the extraordinary significance of China in American minds, even at the top, strangely out of proportion."[24] Of the Allied Cairo Conference in 1943, when Chiang's China was affirmed as one of the "Big Four" allies, Churchill lamented that "the talks of the British and American Staffs were sadly distracted by the Chinese story, which was lengthy, complicated, and minor. . . . All hopes of persuading Chiang and his wife to go and see the Pyramids . . . fell to the ground, with the result that Chinese business occupied first instead of last place at Cairo."[25]

Roosevelt (and his allies-in-the-cause on the right) had several motivations beyond personal bias for interest in China. Secretary of State Cordell Hull recounted in his memoirs that America's war aims included the following: "The first was an effective joint prosecution of the war. The second was the recognition and building up of China as a major power entitled to equal rank with the three big Western Allies, Russia, Britain, and the United States, during and after the war, both for the preparation of a postwar organization and for the establishment of stability and prosperity in the Orient."[26] Hull added, "If there was ever to be stability in the Far East, it had to be assured with China at the center of any arrangement that was made."[27] President Roosevelt also foresaw that in the postwar world there would be "four policemen," and that a united China was the logical holder of the position in East Asia. In May 1942, he told visiting Soviet Foreign Minister Vyacheslav Molotov: "The United States, England, and

Russia and perhaps China should police the world and enforce disarmament by inspection."[28] In sum, a counterweight in the Pacific after the expected fall of Japan, one that would be friendly to the United States, was seen as vital to U.S. policy.

An economic motive was also evident, as there was a growing consensus in American business and government circles that the East would play an important role in absorbing the industrial capacity of the United States. The head of the War Production Board during World War II, Donald Marr Nelson, said in a speech in October 1944:

> For the United States, the war in China means much more than defeat of the common enemy. The stakes of the war are incredibly large, nothing less than the future peace of the world. To this end, it is essential that China emerge from the war as a united democratic country ready and able to cooperate with the rest of the United Nations in a sustained and peaceful expansion of world trade. It's to our advantage and to the advantage of the entire world to see China emerge from this war as the leading industrial nation of the orient, replacing Japan.[29]

The United Nations Relief and Rehabilitation Administration (UNRRA), created in November 1943 and immediately charged with wartime relief, also evolved many plans for the postwar economies of the various nations under its mandate. In numerical terms, China received 72 percent of its aid.[30] An American newspaper profiling the interests of industrialists in China noted that due to its "manpower in abundance . . . there are many phases of China's economy which intrigue the American industrialists. They like to think about bringing the world's greatest manpower reservoir into contact with modern manufacturing equipment. After the war, China's abundant manpower should be able to team up with American productive genius to the advantage of both."[31] Such a vision of China was connected to the previous projection of the importance of China's "mass," from Marco Polo's description of China's "millions" of riches to the economic "Open Door" policy of nineteenth- and early twentieth-century American foreign policy.

In sum, China and America expected much from each other after Pearl Harbor; both, however, were disappointed. Roosevelt, for all his genuine concern about China, agreed with Churchill that the Allies would follow a "Germany First" strategy. The Atlantic theater received most of the resources for war, the Chinese region the least. General Joseph W. Stilwell, commander of American forces in the China-Burma-India theater from 1942 to 1944, put the problem in the graphic terms that earned him the nickname "Vinegar Joe": "Peanut [his own nickname for Chiang Kai-shek] and I are on a raft, with one sandwich between us, and the rescue ship is heading away from the scene."[32] The history of Sino-

American relations during the war and up to the Communist victory, thus, was that both sides were allies pursuing different goals. As the war progressed, this disparity created a tension at home between Roosevelt and the China lovers, who became more and more unhappy with the amount and quality of aid China was receiving.[33] During the civil war to follow, Chiang Kai-shek's American friends would accuse the democratic administrations of either "losing" or betraying China to its enemies; we now can say that such sentiment first gestated in 1942, not 1948.

Almost as if in compensation for this material short shrifting of China, America took up an extensive propaganda campaign on its behalf—one with which the Nationalists cooperated. Madame Chiang, the beautiful, American-educated, eloquent wife of the Nationalist leader, embodied one aspect of this appeal as the proverbial "Miss China" of missionary dreams. But the war years were also a time when, lacking television and the Internet, people received their war images exclusively from the movie screen and magazine and newspaper pages, in the form of both entertainment and news. Combat against an enemy was the best stuff for stories that shaped our visions of China. The advertising copy for one Hollywood film during the war summed up a common theme of casting Japan as the new Yellow Peril: "Jap brutes manhandling helpless women—by official proclamation. See why the villainous Japs have simply got to be exterminated!" In many cases the tie-in to the victimization of China was as heavily drawn as the vilification of the mutual enemy. In the film *China* (1943), starring Alan Ladd, audiences were guaranteed a view of "the screen's No. 1 Man blasting the Japs for what they do to women. . . . Paramount is first to bring you a truly great story of the torture of this valiant land and its defenseless womanhood—and what a Yankee Ladd did about it!"[34]

The contrast between the murdering, raping Japanese and the "valiant" China and its imperiled women was crucial. The "rape" motif continued in the news media. As Harold Isaacs writes, as early as the 1930s "the China war [with Japan] was heavily splashed across the front pages of the American press day after day, and newsreel films, entering upon their heyday as an independent medium of communication, made their vivid impact week after week on a movie-going public."[35] The propaganda campaign was intended to give Americans a positive view of its new ally, and specific contrasts between it and its enemy were common. Second, the campaign tended to negate or ignore unflattering aspects of China, such as the relatively poor performance of its armies and the corruptions and tyrannies of its regime. The U.S. government worked with Hollywood to coordinate the objectives for persuading the American public that, as a manual put it, "China is a great nation, cultured and liberal, with whom, inevitably, [the United States] will be closely bound in the world that is to come."[36]

One example of this principle at work was *The Battle of China*, an installment in the "Why We Fight" documentary series directed by Frank Capra.[37] Capra recounted that George Marshall, then America's head of the Joint Chiefs of Staff, called him personally to solicit his help in making the films. "Now Capra," said Marshall, "I want to nail down with you a plan to make a series of documented, factual information films—the first in our history—that will explain to our boys in the Army why we are fighting, and the principles for which we are fighting."[38] Capra agreed to the project and made, supervised, or cooperated on a variety of "Why We Fight" films that were probably seen by more Americans than any movies in history up to that time. *The Battle of China* (1944) was the sixth in the series (produced by the War Department Signal Corps for the Morale Services Division) and described the "beginning" of the war against totalitarianism, featuring the Mukden incident and Japan's aggression against China. The film visually and through its narrative favorably compared many Chinese ideas, people, and things to those of the West, for example, Sun Yat-Sen and George Washington, and Confucius and Jesus Christ.

The 1940s afford us another way of gauging the effects of such visions of China. Polling science was still in its young adulthood in the 1930s and '40s, and would continue to make considerable and famous errors, but it is one method of trying to understand what ordinary Americans thought about people, events, and issues related to the Middle Kingdom. In July 1941, Luce's *Fortune* magazine contracted the Roper Organization to formally survey attitudes toward China. Of 5,224 participants, only 6 percent answered "more" to the question, "Do you think we should give China more aid, about the same aid, or less aid than we are giving to England?" Then, in September 1941, Roper conducted another survey for *Fortune*, and this time, of the 5,184 participants, 60 percent answered "yes" to the question, "Do you think we should send China all the military supplies that we can spare?" This figure, offered before the attack on Pearl Harbor, suggests strong support for China's war against Japan. It should be recalled that it was the cutting off of oil supplies to Japan that created the rationalization for the Pearl Harbor assault. Was the public willing to risk war by outright military sales? Fortunately—for the purposes of clarification—the surveyors explored the intent of the participants more deeply. Of those who supported sending "all the military supplies that we can spare," the follow-up question was asked, "Do you think we should still do this (send China military supplies), even if it means sending somewhat less aid to England?" Almost 60 percent said "no." Americans, it seemed, were willing to help China, but followed the same "Europe first" strategy that their leaders chose to pursue.

On February 7, 1942, a *Fortune*/Roper poll of 5,196 Americans asked whether the United States and several other nations should be the world's policemen. 44 percent responded "yes," and of these, 2 percent said that the "other country" would be or should be China. In contrast, 73 percent named Great

Britain as our partner in enforcing world peace, and 5 percent named Canada. The Gallup organization also conducted a survey from September 16 to 21, 1942, in which they asked, "After the war, should the United States and China make a permanent military alliance, that is, agree to come to each other's defense immediately if the other is attacked at any future time?" 56 percent of the respondents answered "yes," and only 23 percent "no." Other ambiguities abounded. Americans sympathized with China, but they did not rush to embrace the Chinese in America. The Exclusion Acts were repealed in 1943, but according to a Gallup poll of November the country was split on the issue.[39]

Nevertheless, Americans did feel that the war on the Chinese mainland was part of the war against Japan. When asked what America should do if Japan surrendered but the Japanese forces in China kept fighting, 87 percent of the respondents to a March 1945 Roper poll answered that America should "help clean out the Japanese there." On the other hand, when asked, "What other nations do you think should be our principal partners after the war in preventing disturbances in the Pacific?" China was listed by 43 percent of the respondents, while Great Britain and Russia received 50 percent and 36 percent respectively (the participants were allowed to name more than one country). Then, in May 1945, Roper/*Fortune* conducted a survey of 3,584 respondents that asked people to list the "troublesome problems in this country in the next few years." Only 7 percent cited "our relations with China."

Looking at these responses made during the months before World War II and through the war years, we find a level of ambiguity about China. When asked, Americans were undoubtedly "pro-China." But when pressed, they were not China partisans at the expense of relations with or efforts for America's other allies nor at any great cost (even in materials) to the United States. Also, China was viewed as a potential power—but not a great power, nor even a very important area of issues or troubles in the years to come after the war. This did not mean that the propaganda effort on behalf of China in the United States was a failure. We may say that while most Americans were not definite about what policy toward China should be, they certainly did not support war for China's sake.

In sum, the American public during the war (and for some time afterward) was presented with China visions that contrasted the brave and good Chinese with the barbarous and evil Japanese. The persuasion campaign was not a "top-down" effort, but was put forth by several groups, from the U.S. government itself to many in the press to housewives and businessmen concerned about China. But in these China visions lay a paradox. To some it seemed a contradictory message that the Chinese were helpless "rape" victims but at the same time they were two-fisted allies. Almost all the China images, thus, from news reports to movies, tended to show the Chinese "fighting back" after the initial Japanese onslaught—and winning. But plainly China was losing the war.

3. Communists, "Loss versus Lost," and Contrarians

One observer from the World War II years described the China as seen through Luce publications as a montage of "the gallant losing battles, the brave and clever guerrillas, the millions of determined refugees fleeing west . . . a new country a-building. . . . [A]nd looming over the whole united land [was] the massive figure of the Generalissimo, [Chiang Kai-shek,] his attractive wife only slightly in the background."[40] This "massive figure" did indeed occlude the horizon of China. A March/April 1943 Roper/*Fortune* poll of 5,217 participants asked, "Do you happen to remember the name of the man who . . . is now head of the Chinese government?" 72 percent of the respondents correctly identified Chiang Kai-shek. An academic critic, in a similar vein, wrote in 1943 in the *New York Times Magazine* about "Our Distorted View of China." He claimed that Americans, led by the often "silly" China lovers, to compensate for their guilt at not helping China earlier and even supplying goods to Japan during most of the 1930s, were creating a false image of China that was "a democracy pure and Jeffersonian, its leaders selfless, statesmanlike beings consecrated to the spread of liberty, every peasant guerrilla a boy on the burning deck."[41] These romantic images and narratives composed a vision of China that Hollywood and the Roosevelt administration, as well as the Republican right, approved of and boosted. Yet there were contrarian voices and images. These included some of the most influential government officials, military leaders, and journalists in America. They created other China images that seemed to collide with the Hollywood-Luce-administration line of a "valiant" China "united under Chiang" fighting for "freedom."

One factor in the discrepancy was increasing American familiarity with the "other China" of the Communists. Some background is necessary here, for the timeline of what we knew about the Chinese Communist Party (CCP) is relevant. Within China, the Guomindang (GMD) party was founded in opposition to the rule of the dictator Yuan Shikai, who had come to power after the fall of the Manchus. Named the Chinese Revolutionary Party by Dr. Sun Yat-sen, the GMD set up a stronghold in the city of Canton (modern Guangzhou) and its surrounding areas. After varying fortunes, Sun Yat-sen accepted aid from many different sources, including Soviet advisers. Upon Sun's death in 1925, the Party was split into various factions, all of which nominally pledged support to Sun's successor, Chiang Kai-shek. They had been in alliance with the Communist Party of China, formed in Shanghai in 1921, whose membership had increased to about 18,000 by 1926. Disaster struck for the Communists when Chiang and the Nationalists turned against them, especially those who believed in strict adherence to Soviet policy lines. There were massacres of Communists in Shanghai, Guangzhou, and other cities. Independent Communists such as Mao Zedong began to take control of the party. From October 1934 to October 1935, he and almost one hundred thousand Communist soldiers undertook the "Long March" north, a six thousand-mile journey to escape various encirclement or extermination campaigns by the Nationalists. Between 1936 and 1947, the Communists held their headquarters in Yan'an, a middle-sized town in Shanxi province.

In envisioning the CCP, many Americans—of varying degrees of perception, patriotism, and integrity—diverged in their views. Some saw the Communists as a model of the China the United States should be supporting, with none of the failings of the Nationalists and many virtues that they lacked. Other Americans on the scene (and many more in the United States) perceived the opposite: the CCP consisted of enemies of the free world bent on domination and destruction. Still others fell in the middle: the CCP were not wholly satisfactory, but the United States could work with them. A full history of pre-1949 U.S.-CCP relations is beyond the scope of this study, but there were overall trends worth elucidating. Historically, American relations with the Communists had never been friendly on an official level, although individual Americans made a strong commitment to and friendship with the Communist cause. But America had no role in the first "United Front" between the Nationalists and the CCP that lasted until 1927. During that period, American understanding of the nuances of Chinese government was so poor that it was widely assumed in Washington that since they were working with CCP leaders in the alliance, Dr. Sun Yat-sen and Chiang Kai-shek were loyal Soviet agents.[42] Nor was America prominent in any way in Chiang's later attempt to exterminate the CCP in their battles until 1935; Germany was a bigger patron than America. In this period, American diplomats classified the Communist rebels, who had largely been driven to hinterlands, the countryside, and the mountains, as "so-called Communists"; as late as 1935, Cordell Hull, U.S. Secretary of State, dismissed the CCP as "bandits"—the same term used by Chiang and the GMD.[43]

There were some who were impressed with what they had heard about the CCP, however, including O. Edmund Clubb, then deputy consul general at Hankou. Although later in his career and then in his work as an academic he turned anti-Communist, in the mid-1930s he reported that the Communists embodied a true reforming power within the general corruption of China.[44] No U.S. higher officials, though, seemed to take substantial notice of his view or act upon it. If anything, it was the Japanese war that pushed the American press and public officials to pay more attention to the "northern Reds," especially when it was clear that Chinese public opinion favored a reconciliation of Nationalists and Communists aimed against Japan; the result was a second United Front starting in 1937. Under these circumstances the CCP and the United States began to look at each other from a different perspective. The most significant impetus to find out more about the possibilities of the Communists was the ill feeling among many China service officers and some military leaders toward the Nationalist regime.

From the Communist point of view, there were equally wild swings in visions of America. Although many Chinese (like Mao) had an almost romantic attachment to the idea of America, there was also much bitterness about American foreign policy. In addition, Communist ideology outlined an inevitable and inexorable world struggle between socialism and imperialism, with America being the leading power in the latter camp. As early as 1931, the Communist Party had claimed that the United States was either behind or encouraging attacks on its position by the GMD.[45] Later CCP manifestos identified the Japanese conquest of Manchuria as part of an imperialistic strategy in which America was an ally or instigator.[46] One Party bulletin warned members, "The United States is also an imperialist country as well as the enemy of the Chinese people. To hope that the United States will fight against Japan is to invite a wolf to your house."[47] Even when, on January 28, 1932, Japan landed troops in Shanghai, and in response U.S. ships and Marines were sent to the city, the CCP saw collusion between the two powers: "The main objective of this mobilization is of course to suppress the Chinese revolutionary movement, to attack the Soviets and the Red Army, and to partition China."[48] In 1934, the National Congress of the CCP published a communiqué that attacked FDR's New Deal as a failure and derided the "imperialistic moves of the United States" as made by "imperialist robbers" who "attempt to use the GMD as a weapon to enslave Chinese people and make China an American colony. They are using cotton and wheat credits to help the GMD and attack the Chinese revolution, destroy [the] Chinese economy and enslave the Chinese people."[49] After Pearl Harbor the official CCP position became that the United States was part of an antifascist front.[50]

President Roosevelt himself seemed to be intrigued by the Communists to the point of a sort of distanced romanticism. He sent special agents to seek them out and report to him personally. Marine captain Evans F. Carlson was one such envoy in 1938; he returned by the end of the year brimming with appreciation for what he had observed of the Communists' main fighting force, the Eight Route Army. Back home, his reports and lectures praised the fighting spirit of the CCP troops, the extreme diligence and inventiveness of the commanders, and the bond of camaraderie between both groups. (Carlson borrowed the term "gung ho" from the CCP army and applied it to "working together" to improve training of Marine units.) His books on the subject emphasized the strengths of the CCP army but also the (patriotic) nationalist ethic of the CCP itself. He advised Roosevelt that working with the CCP "would help defeat Japan and secure the friendship of the most dynamic and progressive political group in China."[51]

Reportedly, Roosevelt was a fan of Carlson's writings, comparing them to adventure novels. Roosevelt also, having merchants in the China trade among his ancestors, fancied himself a China expert.[52] Others agreed with the positive assessment of the Communists' utility as potential allies. In 1942, Joseph Stilwell (nominally Chiang's chief of staff as well as commander of U.S. forces in China) requested the use of Chinese Communist troops to help fight the Japanese in Burma.[53] More ambitious plans included landing U.S. troops in northern China and operating in coordination with the CCP armies.[54] The Office of Strategic Services, the predecessor of the CIA, also prepared a plan to "carry out a disownable military program with the Communists."[55] Other doubters about the Nationalist Chinese competence as allies were to be found in high places. A few weeks after the Pearl Harbor attack, Chiang asked the U.S. and British governments for a loan of $500 million as "an advance to an ally fighting against a common enemy."[56] In response, U.S. Ambassador Clarence Gauss cautioned that such support would only assist the "retrogressive, self-seeking, and I fear fickle elements in and intimately associated with the government through the granting of a 'free credit.'"[57] Henry Morgenthau, secretary of the treasury, also advocated limiting Chinese loans, or tying them to efforts on the battlefield, but he was opposed almost entirely by higher officials in the War and State Departments. The House and Senate unanimously passed a bill granting Chinese loans.[58]

Others found themselves doubting the "official line" about China. For example, many Americans (including the president) were fascinated by the early visions of the Communists offered by the most influential China reporter of the era, one who would in later years serve as nearly the sole diplomatic bridge between Communist China and the United States. American journalist Edgar Snow traveled to Yan'an in 1936 and befriended Mao Zedong. His reports and his book *Red Star over China* became sensations; the book was a best seller, often considered the earliest as well as the best book written about the Chinese Communist movement. Snow was one of the select few journalists who tried to see China from the inside.[59] His account of what might today be called the "guerrillas in the mist" was romantic in the literary sense of an exciting narrative about exotic people in a far-off land, but he tried hard to report what he saw and declared that he never represented the Communists to be anything but hardened revolutionaries.[60]

At the same time, however, nobody reading his works could help but contrast the honest, egalitarian, efficient, polite men and women of Yan'an favorably with the corrupt leadership of the GMD. Snow's black-and-white snapshot of a somewhat grim Mao, in dark jacket with a frayed collar, and wearing a cap with a CCP star—lent to him by Snow![61]—before a ragged brick wall and worn door lintel, became an icon of the CCP. As Alain Jaubert notes in his study of the altering of images for propaganda purposes, the picture was often smoothed and "cleaned up" by subsequent Chinese illustrators—some even adding color—to transform the "rebellious, shaggy, unkempt Mao . . . into the stern general of the Chinese revolution."[62]

Another major, if unlikely situated, voice in the CCP's favor sprang from the heart of the Luce empire. Theodore H. "Teddy" White was *Time's* most prolific correspondent in China. Luce, as noted, was an ideologue with a conscience and often hired people who disagreed with him, if they were talented, if for no other reason than that he was genuinely attracted to intellectual ferment. The great publisher's forbearance had limits, however. In the post–World War II period, Luce and White had developed completely different appraisals of the situation in China. In 1946, when White showed Luce the manuscript for a book on China he had written with fellow journalist Annalee Jacoby, the publisher was furious, questioning White's loyalty to him and to *Time*.[63] The book, when it was released under the title *Thunder out of China*, became a best seller and a Book-of-the-Month Club selection.[64] While not "pro-Communist," it did maintain that the CCP was much more successful than the Nationalists in winning the hearts and minds of the Chinese people, and it detailed the failures of the GMD. White did admit, however, in the preface to the 1961 edition of the book, that Luce had been right, at least partly, about the victorious Communists, who had made China "a prison."[65]

On a private level—until his ideas became public through the charges of his enemies—one of the Nationalist regime's most prominent critics was, ironically, the man entrusted with China's survival during the war, the commander of U.S. forces in China, Joseph Stilwell. He was often in conflict with Chiang Kai-shek over many issues. Stillwell argued that China's gargantuan army was less effective than a smaller, better-trained force would be. This allegation was probably true. Each year from 1938 on, the government had conscripted about 1.5 million men, yet the strength of the Chinese army remained at four million, so high was the desertion and attrition rate, not to mention the padding of figures by corrupt generals.[66] The size of the Chinese army was of concern even to those more favorably disposed to Chiang, like General Albert Wedemeyer, who replaced Stilwell after the latter's complete break with Chiang in October 1944. Wedemeyer unsuccessfully tried to reorganize the Chinese army; he did, however, manage to assign U.S. military personnel as advisers to certain Chinese formations.

The issue of the "mass" of China's armies and its aid level was a crucial one. The assistance that America gave China during World War II was judged relatively by different sources. To many Chinese in the GMD regime, it was too little too late, and with too many restrictions. To Americans like Stilwell, U.S. aid was also too little but what there was was so badly spent by the Nationalists as to constitute a great waste. Although few American ground troops served in the China theater, a more effective (and romantic) presence was the American Volunteer group led by Claire Lee Chennault. Chennault was quite different from Stilwell: he liked the Chiangs, married a Chinese woman, and became an enthusiastic booster of the Nationalist cause. Starting with one hundred U.S.-bought P-40 fighter bombers in 1941, Chennault organized three squadrons of "Flying Tigers." They became legends for defeating numerically larger Japanese forces and fought all over the Chinese theater of war. Chennault diverged tactically in this way from Stilwell, who emphasized the ground war.[67]

The division within American ranks was paralleled by an open chasm in Stilwell's relationship with Chiang Kai-shek. The former was "under the Generalissimo in theory rather than fact."[68] The rivalry with Chiang festered to the point where the men were barely on speaking terms. By mid-1943, Harry Hopkins, one of President Roosevelt's top aides and confidants, noted, "The President indicated his very strong dissatisfaction with the way our whole show is running in China. He stated that Stilwell obviously hated the Chinese and that his cablegrams are sarcastic about the Chinese and this feeling is undoubtedly known to the Chinese and the Generalissimo. Furthermore, the President said that it is quite clear the Generalissimo does not like Stilwell."[69] Many agreed with Stilwell and felt that the inflation of Chiang's image in America undermined putting real pressure on him on the battlefront. John Paton Davies, a leading Foreign Service officer (who would fall afoul of McCarthyism after the war) commented wryly in a memorandum to Harry Hopkins on December 31, 1943, "The Generalissimo is probably the only Chinese who shares the popular American misconception that Chiang Kai-shek is China."[70]

Matters came to a head when a crisis erupted in 1944, a year of victory for the Allies on all fronts, but a terrible time for Chinese arms when Japan launched the so-called Ichigo (Operation Number One) offensive in April and overran many Chinese cities and military bases and several airfields used by the Flying Tigers.[71] China by its own counts lost half a million fighting men and countless civilians, and Stilwell was blamed by the government for not releasing more supplies, a situation made worse by the capture of at least ten thousand tons of munitions by the advancing Japanese. From Chiang's view, as he put it in one note to Stilwell, "I do not entertain any doubt that the President is sincere. . . . What has been done is perhaps without his consent or knowledge. As Chief of Staff to me, you are responsible for seeing to it that the promised material is forthcoming."[72] Earlier, Chiang had expressed astonishment at perceived short shrifting, wondering "why the U.S. [had] sent such a pitiful handful of planes and supplies to China."[73] Indeed, for all the paper strength of the army, China hardly constituted a modern nation at war. As one historian has pointed out, "In 1944, China had only about six thousand trucks in the entire country, half of which were out of service due to shortage of parts. This was less than what a small American city would have, or less than the transport allocated to a single U.S. army."[74] For

Stilwell, the lesson was otherwise: the Chinese army's leadership was hapless and guilty of "colossal ignorance and indifference."[75]

Stilwell, as said, peppered his language with denigrations of Chiang and his regime; the man officially known as the "Generalissimo" in the newspapers was called "peanut," "idiot," and "slacker"[76] by America's top military leaders in China.[77] By mid-1944, both antagonists felt that only radical change would avert further disaster. President Roosevelt, following a request by Stilwell and the recommendation of the Joint Chiefs of Staff, cabled Chiang on July 6 and asked him to agree to put Stilwell in command of all Allied forces in China, including both the Nationalists and Communists.[78] Chiang gambled, as he had before and would in the future, that America could not in the end write him off. He refused the request and then demanded that Stilwell be removed.[79] Roosevelt blinked, recalling Stilwell and replacing him with General Albert Wedemeyer, but also charging Ambassador Patrick Hurley to initiate negotiations with the Red Chinese and authorizing a formal fact-finding team to go to Yan'an, an operation that became known as the Dixie Mission.

This initiative was a significant step for the CCP. In anticipation of the Americans' arrival, Mao issued secret instructions that no effort be spared to impress the visitors. The tone of an anticipated reception was highlighted by the lead editorial in Communist newspapers: "Welcome the Comrades of the American Military Observer Group."[80] CCP leaders estimated that a new age of cooperation might be the result, but that no one should overreach by making specific requests for aid.[81] (In this way Mao differed from Chiang, who was always asking for more financial and other assistance.) Among the most important insiders, and the only civilian assigned to the mission, was a Foreign Service officer named John S. Service.[82] No one man better typified the contrarian view of the situation in China within the diplomatic corps nor the controversy that would arise about the "loss" of China to Communism in 1949. Born and raised in China to a family of U.S. missionaries, Service was a true "China hand" with a thorough knowledge of the language, history, and customs of the nation. Serving at various posts, he was assigned to the wartime capital of Chongqing from May 1941 to April 1945. There he met most of the major players in China politics and the military, including representatives of the Communists, at that time in a nominal alliance against Japan with the Nationalists. Zhou Enlai, later to be Communist China's minister for foreign affairs and probably the second most powerful man in the CCP hierarchy, was a frequent conversation partner.

In 1943, Service was assigned to Stilwell's staff. His main duty was to observe, traveling to and writing reports about local conditions in many non-Japanese-occupied parts of China. To other Americans, Service became known as the "authority on Chinese Communism." His opinions were not dispassionate; he felt that the Guomindang regime was corrupt, inept, static, and tyrannical. In contrast, he saw the Communists as leading a dynamic, progressive movement. He argued strongly that the United States should set up some sort of formal post at the Communist headquarters in the hills of Yan'an.

The Dixie Mission confirmed his beliefs, although he cautioned himself "not to be swept off [his] feet" by his first good impression. Mao and the Communist leadership quickly appraised him as their best friend among the American visitors and their most direct channel to President Roosevelt, in contrast to other U.S. military and civilian personnel they suspected of being or knew to be hostile to them. In fact, at the reception dinner, Service was seated next to Mao and was the target of many compliments and thanks from the CCP leadership. At the banquet, Mao asked him to help set up a permanent American consulate at Yan'an. Service's reports were almost wholly laudatory. He said, "My first impression—and those of the rest of the Observer Group—have been to have come into a different country and are meeting a different people. There is undeniably a change in the spirit and atmosphere [with] no beggars, no signs of desperate poverty [and a] YMCA sort of spirit of optimism." The Communist army undoubtedly contrasted favorably with the lethargic legions of Chiang, and thus Service concurred with a War Department report that "the Chinese Communist Regular Army is a young, well-fed, well-clothed, battle-hardened, volunteer force in excellent physical condition with a high level of general intelligence, and very high morale." Other members of the mission, including its director, were also impressed by what they saw; comparing the mud and torpor of Chongqing to the Communist headquarters, one wrote that Yan'an was "like a breath of pure, clean mountain air after a month of London's worst smog." The mission was entertained with poetry readings and dances; most satisfying of all, there was only the merest sign of a Soviet presence in the persons of a few individuals identified to Americans as doctors and journalists from Russia. (This maneuver was, we know now, a purposeful cover-up by Mao.) Service's reports went to the highest level in Washington; those from subsequent visits and contacts with the Communists would also. Service advised extending assistance to the CCP.[83] Many on-site diplomatic and military personnel agreed with him.

Service is to this day the epicenter of controversy about the "loss" of China. The debate has often been framed as a duality: either the Chinese Communists were true Soviet-led Communists and America should have attempted earlier and more effectively to defeat them, or the Communists were mostly native Nationalists driven to a Russia alliance because of American ineptitude and the "lost causes" of the Nationalists both before 1949 and into their years of exile in Taiwan.[84] The shorthand of the debate has been the "lost chance" school, as heralded by Service and his sympathizers, versus the "lost China" school, held by many friends of Nationalist China, conservatives, and anti-Communists (who, we should recall, included many liberals in the 1940s and 1950s).

One of the major "lost chances," and the beginning of the end of friendly U.S.-CCP contacts, was, ironically, the mediation effort charged to Patrick Hurley, President Roosevelt's special envoy, starting September 6, 1944. Hurley, who would later serve as Truman's ambassador to China until abruptly resigning, proved to be singularly unfit to mediate between two antagonists bent on each other's extermination. In the beginning, however, things went well. Hurley met with CCP representatives in October and outlined a bold plan for democratization, national reconciliation, and a joint front against Japan. Mao was initially hopeful that a breakthrough was possible and ordered his forces to cooperate if Americans landed in China in Communist areas. Hurley traveled to Yan'an on November 8 with a plan that had, he thought, been worked out with Chiang and that was amenable to the CCP. After amicable negotiations, a new "Five-Point Draft Agreement" was signed by Mao and Hurley on November 10 to much fanfare. Mao was jubilant, and even wrote to President Roosevelt thanking him and America for their intervention for China's good, and emphasizing the "peaceful" and "democratic" future of China.[85] Returning to Chongqing, however, Hurley met with a brick wall: Chiang refused the terms of the agreement which would have meant real power sharing with the CCP. Mao was angry and felt that Hurley had betrayed him because he had assumed the agreement he had signed with the U.S. diplomat was official and was backed by the United States. In turn, Mao rejected Chiang's counterproposal, which called for the CCP army to be held under Nationalist control.

Deeply distrusting Hurley, Mao tried to contact Washington through other sources, from John Service to Colonel David D. Barrett, head of the U.S. Army Observer Group, on December 8. As Barrett recalled, "Mao's attitude throughout the interview was recalcitrant in the extreme. He was not discourteous to me but several times he flew into a violent rage" when talking of Hurley's betrayal of their agreement and calling Chiang "that turtle's egg."[86] Ominously, Mao warned Barrett that "if the United States would not help us, there is still Great Britain and the Soviet Union."[87] Mao was further enraged at getting no response to his inquiries or to his request to travel to Washington to discuss the situation directly with Roosevelt.[88] Members of the "lost chance" school of Sino-American relations, like Barbara Tuchman in a famous article entitled "What if Mao Had Come to Washington?" argued that this failure was a turning point and the beginning of the downward slide for the chances of good relations between the United States and the Chinese Communists.

Certainly, Hurley tried to sabotage or forbid any extrachannel contacts and seemed to rebuff every attempt Mao made to find some accommodation that would not entail surrender to Chiang. On March 13, 1945, nearing the end of his hopes for American help, Mao told Service, "Chiang Kai-shek is dependent on American help. If he had not had American support, he would have either collapsed before or been forced to change his policies in order to unify the country and gain popular support."[89] A few weeks later, at Hurley's instigation, Service was recalled to Washington, but Mao met with him one last time and made a direct offer to help American forces in any way possible. "Whether or not America extends cooperation to the Communists is, of course, a matter for only Americans to decide. But the Communists see only advantages for the United States—in winning the war as rapidly as possible, in helping the cause of unity and democracy in China, in promoting healthy economic development of China through industrialization based on a solution of the agrarian problem, and in winning the undying friendship of the overwhelming majority of China's people, the peasants and liberals."[90] He even described the CCP as "children without a mother."[91] Hurley's seeming duplicity and the death of Roosevelt further convinced Mao that the light was dimming for any positive relations with America. Events elsewhere had an influence: with Moscow's consent, Allied forces had helped Greece crush its own Communist guerrillas, and Mao on May 8 warned the Seventh National Congress of the CCP that a similar fate might await China's Communists.[92] Within months, Communist publications were referring to Hurley in caustic terms and Mao was making comments about fighting America.

Service's fate seemed to confirm Mao's suspicions. Service was arrested and tried several times, allegedly for passing on secret documents to *Amerasia*, a pro-Communist publication in America. In 1945, Ambassador Hurley resigned, denouncing America's China policy, and later before Congress declared that many American Foreign Service officers were pro-Communist. Senator Joseph McCarthy named Service as one of the top "Communist infiltrators" in the State Department, a "known associate and collaborator with Communists and Pro-Communists . . . consorting with admitted espionage agents."[93] After the Nationalists had escaped to Formosa (Taiwan), some in Chiang's government blamed Service and the Foreign Service group for many of their troubles, detecting a conspiracy to hand China over to Communism.[94] Service, however, was cleared of all charges in the *Amerasia* case in 1945 and then passed a number of subsequent loyalty and security hearings.[95] Other Foreign Service personnel would also face such charges. John Carter Vincent, another China hand, was saved in 1947 from charges that he undermined U.S. policy in China only by the intervention of Secretary of State Dean Acheson.[96]

These circumstances did not bring about the end of American flirtations with the CPP. Another major initiative came in response to Hurley's resignation and his charges against the administration. The Marshall Mission was the name given to the extended effort between December 20, 1945, and January 8, 1947, by George C. Marshall and others to broker a postwar peace between the Nationalist government and the Communist forces in China. President Truman correctly predicted that Marshall's immense prestige would act as a temporary firebreak in criticism of his China policy, and indeed the mission was lauded even among the strongly pro-China press like *Time* magazine. As with other attempts, this mission seemed at the outset to be heading toward success. On January 10, a cease-fire accord was worked out. A series of other agreements on issues such as the structure of a constitutional government and a troop reduction on both sides followed. Shortly afterward, however, the talks stalled on certain irresolvable points. Indeed, Marshall would have succeeded only through some miracle, for before and during his stay the GMD was accelerating political repression and superrightist elements were engaging in political assassinations in areas they controlled. Elections in 1947 that led to a new parliament seating in January 1948 yielded no democratic opposition of "middles" but rather resulted in an even stronger right-wing power base for the GMD. The Communists refused to dissolve their million-plus-man army and integrate it into a national Chinese army under Guomindang officers. The Guomindang, in turn, were not willing to cede any political power to the Communists except some token seats in the legislature and certain unimportant ministries. The truce cracked as both sides maneuvered their armies for battle, and when the cease-fire expired on June 30, war broke out. Negotiations continued, but the intransigence of the antagonists could not be breached; Marshall himself had finally given up on January 8, 1947, returning to the United States to become secretary of state.

Part of Marshall's problem was that no one took his "big stick" seriously. For example, the United States embargoed arms and munitions to China from August 1946 to May 1947. This action did not affect Nationalist policy because Chiang was convinced that he could win with arms in hand or obtain arms from other sources. The Communists themselves did not take the embargo as a real check on Chiang's ambition to destroy them. For his part, Marshall sincerely believed that Communist control of China would hurt American interests. But at the same time he felt that if Chiang Kai-shek's forces won, China would be a commitment for the United States "of uncertain magnitude and indefinite duration."[97] Moreover, the flaws of the GMD were so serious that only immense aid from the United States would help, and he was unwilling to recommend it.[98] Much in the same vein as Mao's warning after the failure of the Hurley mission a few years earlier, Zhou Enlai cautioned Marshall in his last days in China that the Soviet Union could do for the CCP what America would not.[99]

In any case, 1948 was a year of disaster for the Nationalist cause. In the period 1945 to 1947, Chiang's armies won many victories, the most symbolic of which was the conquest of Yan'an. But the CCP forces, infiltrating the Nationalists' rear areas, destroyed 10,250 miles of railroad track, and the GMD government managed to repair only about a third of it.[100] CCP spies were also very successful in keeping their leaders apprised of government plans and moves.[101] In 1948, city after city and province after province fell to the Communists. The most decisive battle was that of Huai-Hai, fought between November 6, 1948, and January 10, 1949. By the end of the struggle, actually several major battles, most of China north of the Yangtze River was under CCP control. Worse, Chiang Kai-shek saw his best remaining troops captured, killed, or dispersed. The tactics used in the battle were indicative of how the Communists won as well as how the Nationalists lost. While the Nationalists tended to concentrate their forces on holding strong points, especially dense population areas like cities, the Communists would spread out, seizing the countryside and routes of communication and supply, focusing their attacks on Guomindang reinforcements. At the end of the final battle between January 6 and 10, 1949, about half a million Nationalist troops were killed or captured. In response, Chiang Kai-shek officially resigned as president of the Republic of China on January 1, 1949, and the way was opened for the Communists to cross the Yangtze into south China. Preparations began for a flight of the GMD to the island of Formosa (Taiwan).

At this point it became clear that U.S. foreign policy's primary concern was when and how to abandon Chiang and how quickly to recognize the CCP as the victors. Such a dilemma was very much driven and circumscribed by domestic politics. An aside is needed here to note that views of China during the civil war and after the Communist takeover in 1949 tended to fall into certain polarities—and still do. They may be roughly divided into the "loss of China" versus "lost chance in China" categories mentioned above. In the first group were those scholars, diplomats, politicians, journalists, and others who argued that America deserved most of the blame for China's "going Communist" because the U.S. administration failed to fully support the GMD and to recognize the scope of the Communist threat, especially that the Red Chinese were in the end no more than fronts for Soviet Communist aggressions. This view was held not by a few specialists only but was a popular notion in the United States. Journalist Joseph Alsop, a longtime friend of the Chiangs, was furious at the Truman administration's contention that the Nationalist cause was doomed and unsalvageable. "If you have kicked a drowning man briskly in the face as he sank for the second and third times," Alsop claimed, "you cannot later explain that he was doomed anyway because he was such a bad swimmer."[102] In an early, massive, and highly influential study, Tang Tsou, a Chinese American historian, argued, "Any policy based upon a refusal to intervene militarily in Chinese civil war would have been doomed to failure."[103]

The "China lobby" now came into public prominence as a group of Americans and Chinese who were vociferous supporters of Chiang Kai-shek and the Nationalist regime.[104] Among the "members" of the lobby were such disparate figures as Senator Joseph McCarthy (R-Wisc.), *Time/Life* publisher Henry R. Luce, and influential journalist George E. Sokolsky. The wider circle of the China lobby included General Claire C. Chennault, head of the Flying Tiger squadron in World War II and friend of Chiang, who defended the Generalissimo to the point of attacking anti-Chiang American officers like Joseph Stilwell. Support for Chiang came from many quarters, including liberals and Democrats, and evidence from public opinion data suggests that he and especially his wife were genuinely popular in the United States.[105] Among the press barons considered pro-Nationalist were, besides Luce, Roy Howard (Scripps-Howard), William Randolph Hearst (Hearst Press), Robert R. McCormick (*Chicago Tribune*), William Loeb (*New Hampshire Morning Union*), and many leading politicians (a number of whom were former missionaries).[106] By one count the "China bloc" on the Hill included one-quarter of the Senate and one-eighth of the Congress.[107] The leader on the Senate side was California senator William Knowland, nicknamed the "Senator from Formosa."

The tide of sentiment and scholarship tended to turn in the late 1960s and the early 1970s, especially in reaction to Nixon's visit to China in 1972. John Service and other "old China hands" began to publish their papers, newly declassified, and make a case for "a lost chance in China."[108] Among their theses was that there were several opportunities going back to the 1930s for good Chinese Communist-American relations, that America had been the chief saboteur of these possibilities, and that the latter failed to perceive the many irreparable flaws in the Nationalist regime that would have led inevitably to their defeat. A key observation was that the Chinese Communists did not necessarily have to end up in the Soviet camp, but could have been, if not pro-American, "neutral" or at least not hostile toward America. As Service outlined it, "If the United States had been able in 1945 to shed some of its illusions about China, to understand what was happening in that country, and to adopt a realistic policy in America's own interests, Korea and Vietnam would probably never have happened. . . . We would not still be confronted with an unsolvable Taiwan problem . . . because we would not have 'lost' China."[109] New evidence, obtained by the opening of archives in America, China, and the former Soviet Union, suggests that the matter is complicated. The CCP were committed Communists, but not committed to being close Soviet allies. At the same time, the Nationalists were quite willing to work with the Soviets, and often ran a double game, playing up fears in America of a Soviet China if they lost the war to the CCP. It seems both Chinese sides attempted the old strategy of the emperors to "use the barbarian to control the barbarian."[110] In short, it seems reasonable to say now that at best Service was less a player than a well-played figure in Sino-American relations.[111]

The "loss of China" and the "lost chance in China" schools of thought both assume that America was the decisive factor in China's wars, and fail to admit that the Chinese largely controlled their own destiny. This latter point is underlined by the fact that both America via the Marshall Mission and the Soviet

Union through a similar follow-up mission tried hard to broker a deal between the warring parties. Indeed, the Soviet Union was painfully scrupulous in keeping good relations with Chiang Kai-shek, signing several treaties and cooperating in troop movements and diplomatic initiatives and consistently pressuring the Communists to find an accommodation late into the civil war. Stalin, as of 1947, was apparently fearful of a Western-Soviet confrontation and had no interest in making China the tinderbox of another global war. At the Yalta Conference in 1944, Stalin had made a promise to Roosevelt that he would not interfere in the Chinese civil war and he accepted the view that at least the southern part of China would be within the Western sphere of influence.

Stalin's secondary motivation was the experience of the fear in 1941 of the USSR being attacked in the east by Japan; he therefore did not want China unified under any one power, even the Communists. As late as February 1948, the American consul general in Beijing, O. Edmund Clubb, was reporting that the Soviets were working on a peace deal and putting pressure on the CCP to accept it.[112] In April 1948, Stalin even refused to allow Mao to travel to Russia for a meeting. Another crucial indicator of this Soviet neutrality was that significant CCP military victories were hardly mentioned in the Soviet press. Finally, in February 1949, when the Nationalist government abandoned Nanjing and fled to Guangzhou, the Soviet ambassador moved with them, to the frustration of the CCP leadership. This sort of behavior led Mao to comment, "The Chinese revolution achieved its victory against the will of Stalin."[113] In any case, a series of statements by Zhou Enlai and Mao in April and June of 1949 offered the "five principles of peaceful coexistence" between a Red China and "all foreign countries on the basis of equality, mutual benefit, and mutual respect for territorial integrity and sovereignty" as well as self-reliance, and the resolution of issues peacefully "on just grounds, to our advantage and with restraint."[114] The door to some sort of accommodation seemed to be still open.

American diplomatic flirtations continued with the Chinese Communists, including talks between John Stuart, the American ambassador, and Huang Hua, a leading Chinese diplomat. These talks were the subject of another "loss versus lost" debate, more so since Huang actually invited Stuart to visit Beijing for more extensive negotiations but Secretary of State Dean Acheson forbade the trip.[115]

One of the moderate proponents of the "lost chance" school of thought, Warren Cohen, argued that "Acheson's goal was to reach an accommodation with the People's Republic under the most advantageous terms for the U.S. He hoped to encourage the Beijing regime to distance itself from the Soviet Union and to recognize the importance of its historic ties with the West . . . even the coming of war in Asia did not preclude accommodation with the People's Republic. . . . He might well have succeeded had it not been for the decision to send U.N. forces across the 38th parallel."[116] Similarly, Nancy Tucker argues that

Acheson had no faith in or fidelity toward Nationalists and rather wanted "to find ways to reconcile United States interests with those of the incoming government of China."[117] It is clear that the Chinese leadership believed then—and still does—that the failure to reach some type of formal relations was on the American side.[118] There is no question that Mao and his diplomacy teams blamed the United States for the failure of a rapprochement in the 1940s.[119]

At the same time, new native Chinese scholarship has admitted that the "Yan'an-insulated" CCP leadership had little understanding of the why and wherefore of international diplomacy and its protocols.[120] They also lacked a full appreciation of the uncertainty Truman and his advisers felt about their ally Chiang.[121] The CCP's efforts to appease Moscow led them to "freeze out" even those Americans disposed in their favor. For example, it is likely that Edgar Snow was declared "persona non grata" in the People's Republic precisely because he emphasized in his writing the independence of the CCP from Moscow.[122] (Another example of Mao apparently thinking that the American press signified the policies of the government.) The party's determination to "make a fresh start" in foreign policy entailed a slow, sensitive, and careful process toward making amends with Washington or, as Mao himself put it, "As for the question of the recognition of our country by the imperialist countries, we should not be in a hurry to solve it now and need not be in a hurry to solve it even for a fairly long period after country-wide victory."[123] In many other statements and memoranda throughout 1949 and 1950, the CCP leaders accorded good relations with America as not of vital interest to China in the short term.[124] Perhaps they were being practical as well as political, for they must have sensed that their litmus test of American good intentions—that Americans renounce "unequal treaties" and basically apologize for past conduct in China—was highly unlikely. Yuan Ming, deputy director of the Institute of International Relations at Beijing University, has argued, "The largest failure of American policy was the lack of understanding of the most basic inclinations of the leaders in Beijing—that is, the United States confronted a group who were not only Communists but who were also determined to rid their country of all its national humiliations since 1840. . . . Another misunderstanding in Washington related to its belief that the CCP was just a tool of Moscow."[125] In addition, China's "mass" once again played a role in its conduct and that of others toward it. Mao believed that China was big enough to rebuild itself without Western aid; of course, Westerners saw in that vastness evidence of his need for aid. In sum, in Washington and in Communist China, rapprochement had few friends and no lucky breaks.

The consul general at Beijing, Edmund Clubb, reported in 1949 that he saw an opportunity for some sort of deal between the new regime and the United States.[126] He wrote, "Diplomatic relations does not imply judgment of its domestic politics, and however unpleasant it seems [the] USA can hope to play an effective pacific role in contemporary China, and perhaps manipulate [the] situation to its political advantage, only through [the] establishment channel for direct dealing with [the] present regime."[127] By late December, Clubb was even more insistent that the United States should make its move, for "it seems . . . probable [that] if American government takes, with or without participation from Congress, the position that recognition definitely [is] not to be extended, Communists will in due course call a halt to even present limited functioning Consulates and possibly call for their closure and our departure from China."

Clubb believed a continued U.S. presence in China was vital, and he dismissed the two standard objections to according recognition. He believed the Communists were in control of the nation and questioned whether they would be any less willing to live up to their international obligations "than the east European satellite states."[128] Then in January 1950, despite the Communist confiscation of American consular buildings, Clubb, although admitting the situation was "problematical," continued to believe that the Communist government would respond favorably if the United States assured recognition.[129] His superior disagreed. Acheson declared, "The seizure of the Beijing consulate made it clear that the Chinese were not interested in U.S. recognition."[130]

Another crucial event—almost ignored in the West—occurred in late October 1949, when a large force of People's Liberation Army soldiers tried to seize Jinmen (at the time known as Quemoy), one of the small offshore islands in the Taiwan Straits, a stepping-stone to Formosa. The GMD was ready, however, and proved to have superior naval craft, artillery, and fighter bombers that pounded the invaders and then cut them off from resupply, reinforcement, or evacuation. Almost ten thousand PLA troops were captured or killed.[131] It was a huge defeat in terms of men and materials, but also in morale. Mao felt the entire PLA had been chastened. The battle was also a great hypothetical of history. The PLA's plans for a cross-strait invasion were put off for a long while, never to be seriously replayed to this day. The battle also taught the PLA to more carefully gather intelligence about its foes and not just trust in Communist "spirit" to achieve victory—a lesson that would pay off in Korea. In addition, the American equipment that proved the decisive factor in the battle was taken as a symbol that the Taiwan question should be viewed in light of the international struggle between the "imperialist" and the "socialist" camps.

In such an atmosphere, the failure of rapprochement talks marked the closing of the curtain on yet one more "chance" for Sino-American rapprochement. On June 30, 1949, Mao made the statement that would seem to define the next decade of foreign policy: "Externally, unite in a common struggle with the peoples of all countries. That is, ally ourselves with the Soviet Union, with the People's Democratic countries, and with the proletariat and the broad masses of the people in all other countries, and form an international united front. . . . We must lean to one side." Mao's motives for this stance have long been debated, but they included a wish to be in good stead with Stalin, anger at American actions, and basic Communist ideology that perceived inevitable conflict with imperialism.[132] But there was another factor, a historical one. Westerners, from Marco Polo onward, had a history of viewing China as a country that needed to be "opened"; the Chinese, whatever their politics, felt anger and humiliation at foreign impositions. Unequal treaties, forced rights: these were associated with "diplomacy" for many in the CCP leadership. As Mao commented, "We are willing to establish diplomatic relations with all countries on the principle of equality, but the imperialists, who have always been hostile to the Chinese people, will definitely not be in a hurry to treat us as equals. [If] the imperialist countries do not change their hostile attitude, we should not grant them legal status in China."[133]

Here the "lost chance" theory becomes dubious. The historical reader, weighing the great debate on China's fate, will note that the CCP leadership said many things and did many things that would support both a chance for good relations and no chance for any relations. Perhaps this "confusion" was not mixed signals nor a shifting mind-set but deception: people selectively perceiving in 1949 and today what they wanted to read in Mao's or Zhou's declarations. New China scholars are almost certainly correct in their assertions that Sino-American relations were driven largely by China and that Mao, although he was to misread America in many ways, did recognize that the "imperialist" enemy could be put off by the right words and phrases. According to this scenario, Mao's anti-American positions were fundamental to his policies and goals, and all talk to the contrary was smokescreens and mirrors. His basic goal was to neutralize the possibility of America's saving Chiang. As Michael W. Sheng has written,

> During both the civil war and the Korean War, Mao's "credibility concern" had two dimensions: to raise morale within the revolutionary camp, and to frustrate the morale of the enemies to force them to retreat (*zhinan ertui*). Therefore, by announcing the "hand-off" policy, the Truman Administration conveyed a sense of hopelessness, and Mao's "tiger spirit" was thus encouraged, which made him [Mao] more inclined to adopt a hard-line policy in Korea when the crisis erupted there.[134]

It was one case among many of Mao's being entranced by his own propaganda. His "diplomacy" was in part simply "disinformation to confuse the Americans so that a possible U.S. direct intervention against the CCP's takeover could be avoided or delayed."[135] In such a light, America's policy of containing Mao rather than trying to win him over becomes more historically justified.[136]

The other problem with a "vision" of a possible Sino-American relationship was that the American political situation seemed to preclude it as much as any Marxist theory. The American side had their own stone wall. Gene David McLean cites ample evidence that Dean Acheson was making overt and covert anti-Communist moves that would hardly signal even the smallest willingness to make a deal until October 1949 when the CCP declared a new "People's Republic of China" as sole government of the country.[137] That after October he may have been more interested in accommodation was not of much import, since by then the opportunity had passed. Michael Schaller also argues that practically every key decision-maker at the State Department was a strong anti-Communist, and even "utmost moderates had lost nearly all faith in rapprochement with the Chinese Communists by the spring of 1950."[138]

The options open to American policymakers by the 1948–1949 watershed were limited, even if they thought more like Service than Hurley. The Truman Doctrine, proclaimed on March 12, 1947, and Marshall's European Recovery Program were attempts to draw a line against further Communist expansion, but with an emphasis on Europe. Nevertheless, the pro-China bloc in Congress and in the press provided a check on any inclination for Truman to completely write off the Nationalists. Truman's China policies came under heavy fire in relation to what America was doing for Europe.[139] Indeed, it can be said to be one of the negative costs of the Marshall Plan that China Firsters and the China bloc were able to successfully argue, especially in Congressional hearings and in editorials, "If Greece, why not China?" Consistently, military and economic aid to Europe was seen as linked to the same for the Far East. This linkage frustrated the father of containment, George Kennan, because he did not feel that the United States had the resources to fight Communism over every inch of the world, either economically or militarily. Kennan said, "If I thought for a moment that the precedent of Greece and Turkey obliged us to try to do the same thing in China, I would throw up my hands and say we had better have a whole new approach to the affairs of the world."[140]

But Truman needed Congress, so he signed the Republican-controlled Congress's China Aid Act of 1948, which included $275 million for economic and agricultural purposes and $25 million for military purchases.[141] (Mao denounced the bill and said the U.S. action was "the fundamental reason for the outbreak and escalation of China's civil war."[142]) It would be hard to conceive that this relatively small amount of money would achieve its stated goal "to maintain the genuine independence and administrative integrity of China and to sustain and strengthen the principles of individual liberty and free institutions in China."[143] George Marshall's director for Far Eastern affairs, Walton Butterworth, admitted that "the aid bill arose out of the fact that Senator Vandenberg [who wanted to work out a compromise] let it be known that the chances of getting through the Marshall Plan would be greatly improved if there was a China Aid program."[144] O. Edmund Clubb also argued that "President Truman felt it necessary to mollify the opposition to some extent by adopting a Cold War stance, thus to enable him to put across the Marshall Plan for Europe."[145] April 1949 brought $54 million in aid to the Nationalists and a little later another $75 million.

The support came with an anticipation of Chiang's collapse. The consensus of Foreign Service officers in China was that the economic aid was not helping and could do neither the Nationalists nor U.S. interests any good.[146] In a National Security Council meeting in March 1948, China's position was reviewed, and all agreed that the Soviet Union had designs to control China and then move onward into Southeast Asia.[147] Several possible responses could follow. The first option would be to cut all aid to Nationalist China. No one denied that this would lead to the immediate collapse of the Chinese government and a Communist victory. Another proposal from the Defense Department, with the support of Secretary of Defense James Forrestal, was for "limited economic and military assistance to the National Government of China." This suggestion was opposed by Marshall and his group, who favored only limited economic, but not military, assistance. Most were convinced that the Nationalist position was too weak to benefit from any level of U.S. military intervention.[148] No one supported all-out military intervention in China. Dean Acheson, secretary of state at that time, told a meeting of pro-Nationalist Republican congressmen on February 7, 1949, "[W]hen a great tree [Republican China] falls in the forest one cannot see the extent of the damage until the dust settles."[149] In the autumn, the U.S. Department of State convened a conference on East Asia and concluded that the complete collapse of the GMD was only months away and that the United States would end up recognizing the Communists "fairly soon."[150]

A final context for understanding the history of the vision of China in 1949 was the progress of the Cold War, which unfortunately caused the United States to add a final blunder to the China affair. The year 1949 offered some signs for optimism about America's foreign policy. Early Cold War battles went well: the Berlin blockade petered out in May, and the anti-Communists won the Greek civil war.[151] Some hopes also arose from intelligence (and press reports) that there was growing Sino-Soviet tension, of which many in the U.S. administration wished to take advantage.[152] As early as 1947, Kennan increasingly believed that America should focus its efforts on encouraging Nationalist and anti-Soviet elements within the CCP.[153] He argued to Marshall in November 1947: "[There was] no convincing evidence that, even should the Chinese National Government collapse, the Communists could in the foreseeable future assume effective authority over all China and at the same time remain seriously susceptible to Soviet guidance or control in international affairs. Thus, while a collapse of the National Government would be deplorable, it probably would not be a catastrophe for American interests in China."[154] China was perceived to be "a vast poorhouse" with only a few regions like Manchuria of actual economic worth to the Soviets.[155]

Among the American public, there was also no great support for an aggressive China policy. The Gallup organization, in a poll dated from the end of November to December 1, 1948, asked people whether they had heard or read about "the civil war in China." 79 percent said they had. Of that number, when asked "Would you favor or oppose sending Chiang Kai-shek's Nationalist government about five billion dollars worth of goods and military supplies in the next year to try to keep China from going Communistic [sic]?" only 32 percent favored the proposal, while 34 percent opposed it. The same 79 percent who had heard about the war were asked about its "status." Almost a third of the sample described the situation as some variation of "Communists gaining ground; Nationalists losing; desperate situation for Chinese government." 16 percent said, "Chinese people suffering; no food, no clothing." (The rest of the answers were general, confused, or consisted of "don't know.") Then, when those who reported hearing of the civil war were asked, "Do you think the Chinese Communists take their orders from Moscow or not?" 51 percent answered, "They do"; 10 percent said, "They do not," and 18 percent offered no opinion. Another question asked, "What is the first problem you would like to see President Truman and the new Congress take up?" China was mentioned by only 2 percent of the sample. In sum, Americans opposed the notion of China "going Communist," but were not really convinced the process could be halted. This general sentiment undermined the China lobby only slightly, for even if just a small part of the electorate feels strongly on an issue it may disproportionately effect elections. Truman may have known that America as a whole was not obsessed with China, but enough influential people and some voters were, and many others saw China as an important symbol of foreign affairs. This fact limited Truman's actions as much as any groundswell of public opinion.

The year 1949, then, was one of decision for many policymakers throughout the world. For their part, the American news readers and viewers would have had several competing "visions" of China gleaned from many sources, from popular films to presidential speeches. What then did they actually see in the nation's leading newsmagazine?

4. 1949: The Pictures

The year 1949 was one of crushed hopes and dreams for Henry Luce. A lover of China, he hoped that China would play an integral role in the "American century" to come that he had envisoned during World War II. To see the beloved second home-

land of his parents overrun by the Red hordes of Communism did not shake Luce's worldview but profoundly challenged it. According to his friends and coworkers, Luce reacted with a combination of uncertainty and maturity. He still cared about China; he still felt that Communism was no answer to China's many problems. However, the publisher felt that, in a sense, Chiang Kai-shek had failed both him and the Chinese people. But he himself wondered whether Nationalism was the absolute good versus Communism being the absolute evil. Such an ambiguity is well reflected in the imagery and verbal descriptions of the tumultuous year of the death of the Nationalist regime on the mainland and the birth of the People's Republic of China.

Time*'s vision of the Middle Kingdom for 1949 is alternately gloomy (the bamboo curtain is descending and the Chinese people will be crushed), hopeful (some sort of accommodation will be worked out by Communists and nationalists), fantastic (a new vigorous Nationalist group will emerge to replace Chiang and to fight Communism), determined and defiant (the struggle will continue for Taiwan), and even liberal and open (the Communists are a fact, we must deal with them). Such views did not spring from the mind of only one man. They reflected the stops and starts of American policy toward the events in China that increasingly were not able to be controlled by Washington, if they ever were. Regardless of who was correct in the perennial "who lost China" debate, and whether or not normal relations could have been established between the United States and the PRC, it is clear that while diplomatic windows were closing, Henry Luce's* Time *magazine left doors open for accommodation. Luce as a great publisher esteemed power, and he respected success as a great capitalist. One gets the feeling that, if there had been some sort of rapprochement between the United States and mainland China in late 1949, the Lucepress would not have worked very hard against it.*

To gain an impression of the pictures of China that composed its overall visions in the year 1949 we first begin with a quantitative overview. *Time* published forty-six China-related pictures, including one front cover. The most frequent subjects of news images were politicians and military officials. *Time* devoted about 67 percent of its China pictures to the civil war between Nationalists and Communists. It is worth noting that, according to a previous study, this adds up to less China coverage for 1949 than that of *Time*'s rival *Newsweek.*[156] But this overview does not capture the nuance of the visions of China in the individual images, their context within the clusters of associated words and other pictures on the page, or the progression of images from week to week in relation to news events. The last reveals a more ambiguous view of China.

To begin, on January 10 there are two China pictures in *Time.* The first is of a "Junk jam in Soochow Creek, Shanghai" whose caption reads, "To ease the strain, a soft line." The accompanying stories in the foreign news section explain that "Communist boss Mao Zedong was striving to ease the strain of revolution by talking of moderation." Citations of incidents, quotes, and evidence are deployed to note that the "Reds" were trying to avoid excesses and, as one Communist soldier is quoted, "not all Americans are against us." The story also emphasizes that attacks on "American imperialism" seemed ritual and rote. In contrast, a second story—illustrated by a picture of one of the Nationalist leaders who urged negotiations between the sides—describes the eclipse of Chiang Kai-shek. The latter had been asked to resign or step down by many in the GMD; the story reports, "In spite of Chiang's tough talk, it looked last week as if the Nanjing government might be willing to make a deal." A possibility was that "the Gimo might retire to Formosa."

In all, then, the first visual and verbal descriptions of the situation in China in 1949, just as the Nationalist forces were being smashed in the northern battles, was of ambiguity and optimism about the CCP. That this would later be shown to be journalistic overoptimism does not undercut the understandable hope that some accommodation and rapprochement might be worked out with the Communists. The problem was, as noted, that these press reports were following the line of most government observers, and the reporting itself created an atmosphere of suspicion between the Russians and the CCP that led to a course of "freezing out" Americans. Visually and verbally, such reports from China may have been a case of a self-denying prophecy, for American news elites as well as the public.

On January 17, a *Time* photograph showed "Nationalist troops near the front" walking down a dusty road. The accompanying story, "When Headlines Cry Peace," described a growing movement within Nationalist areas to make peace or switch allegiance to the Communists. This story correctly alleged, however, that "the high-riding Reds seemed to have no thought of making peace except on their own terms." It is also noted that, in the besieged town of Tientsin, in north China, radio propaganda about "running dogs of American imperialism" preceded Communist artillery bombardment. In the same vein, a picture in the January 24 issue of *Time* showed a close-up of a young girl who had fallen asleep begging on a street corner with a younger child strapped to her back; both of them have heads drooped in utter weariness, the girl's tin cup empty. The picture is explicitly described as a metonym for the situation in the country: "As she slept clutching a beggar's cup in her hand, the picture she made was a picture of Nationalist China itself. Last week the bad news that poured into Chiang Kai-shek's tottering capital would have exhausted even the strongest." The story detailed the Communist rejection of peace terms and the continued collapse of Nationalist armies on all fronts.

The issue of January 31 contained three images; all reveal a strong political intent on *Time*'s part. First, a Disneyesque map shows a bright red liquid tidal wave of Communist conquest pouring toward the Yangtze River (which bisects central China east to west); arrows indicate the retreat of the Republican army toward the southwestern inland city of Kweilin, the movement of the capital to Guangzhou, and Chiang Kai-shek "retiring" to the coastal city of Fenghwa (on the way to Formosa). The accompanying story, pivoting on the fate of China's mass and entitled "What Can We Do?," describes the major news event that, "defeated and helpless, Chiang Kai-shek, for twenty-two years the dominant figure in China, stepped down last week. His retirement symbolized one of the great shifts in the twentieth century's turbulent history: some 460 million Chinese, a quarter of the human race, were passing under the domination of Communism." Later in the story, the global significance of the event is underscored: "Mao's victory made a major change in the political and strategic world picture on the western shore of the Pacific. From Bering Strait to the Gulf of Tonkin, Communism was now the major force. The western world merely held sentinel positions in Japan, the Philippines, and Indonesia. . . . Not since Hitler had stood on the French coast looking west across the Atlantic had the danger been so great." (It should be noted that South Korea—for which America would fight within a year—is *not* itemized in this list.) Although Chiang is portrayed as a defeated man, there are many details provided of pathos and sympathy, such as his attendance at "Christian services with high school students" near the grave of his mother. Later, a picture of a smiling Mao is shown with the caption, "He had what he wanted." The final image of the week seems incongruous: "Chiang and cheering Nationalist troops" shows the Generalissimo walking between two rows of saluting soldiers. The picture may be meant as a last tribute to an old friend, subcaptioned with a whiff of martyrdom. Such an interpretation best fits with *Time*'s editorial stance, although certainly Chiang's enemies might view the same picture as a representation of ridicule.

In the week of February 7, 1949, *Time* magazine signals the beginning of a new era in China with a potent visual metaphor: "China's Mao Zedong" is featured on the cover of the issue. It is an extreme close-up brush painting, in *Time*'s quasi-Norman Rockwell style. Mao is in a tight-collared "Communist jacket"; his eyes look off in the standard "far horizon" gaze of the visionary leader. He is definitely not an attractive figure, however. His facial lines are deeply etched; thin lip, scanty chin hair, and a half-grin/sneer give him a distinctly sinister appearance that, while not quite Fu Manchu, looks nothing like the airbrushed quality of Chiang Kai-shek's portraits in the '30s and during the war years. The subcaption leaves no confusion as to which direction *Time* estimates the new regime to have taken, and what the reader-viewer should conclude about the subject: "The Communist boss learned tyranny as a boy." The story features in detail the many military advances of the Communists and surrenders of the Nationalist position all over China. Along with some other pictures, what follows is a minibiography of Mao. It is not a wholesale condemnation, but it does describe a man who was part "political commissar" but who had the "humor, strength, and often the manner of a Chinese peasant." There are hints of a hope of rapprochement. One image shows a long-ago celebration at Yan'an of the Allied invasion of Normandy.

The article ends with the question, "What kind of master will Mao be to China?" The story mentions that some have thought the members of the CCP to be not pure Communists but rather "agrarian reformers"—this may be a jibe at journalists like Edgar Snow and former Luce China hand Teddy Snow—but that the evidence is that Mao will be a dictatorial Communist like all others. Of particular interest is that the story finds that initial cordiality in the areas that were conquered ("nod head") is almost always followed by increasing intransigence ("shakehead") and then by brutal control ("chophead"). The article concludes with a paean to the Chinese people, noting that they have endured many conquerors in the past, including the Tatars. In the years to come, *Time* will resurrect this romanticized theme.

The pictures of China in *Time* magazine's following few weeks focus on the transition from GMD to CCP rule in the major cities, from Guangdong to Beijing. In the issue of April 25, there is an image of Chiang, in casual clothes and leaning slightly on a cane, in the city of Fenghua, detailing more of the retreat of the Nationalists. In a rather curious use of an adverb, the story describes Chiang's decision to move $300 million of the Nationalist gold and cash reserves from Nanjing and Shanghai to "safer vaults in Formosa and south China" as being made "thoughtfully." In following issues, maps, portraits, and other pictures document yet more territory falling to the Communists. One piece in the issue of May 2 notes how Communist soldiers invaded the U.S. embassy in Nanjing. On May 16, there is a picture of "an execution in Shanghai" that shows Nationalist troops standing over the dead, bleeding bodies of prisoners who had "plotted to overthrow the government." Personalities are also introduced visually, for example, Communist generals and surrendering Nationalist ones. *Time* continues to follow Chiang's course toward Formosa. In all, news from China is rather sparse.

There is a discussion in the June 27 issue, accompanied by a map and a portrait, of a Nationalist warlord in northwestern China who "stood a fighting chance to hold the northwest for some time to come." The general in question, Ma Pu Fang, is described as "a pillar of anti-Communist strength" and a "high-handed but benevolent despot." The issue of August 15 focuses on the domestic politics of China. A portrait of Secretary of State Dean Acheson is subcaptioned "In a moment of anguish." A cartoon shows a man whose head is in the form of a globe perusing a newspaper whose headline reads, "Reds Lose Out in Europe"; he sits on a barrel of presumed dynamite, complete with burning fuse, labeled "Asia." This caricature, as noted above, is a reference to the months when Cold

War events—the failure of the Berlin blockage, the victory of the anti-Communists in Greece, the rejection of Stalinism in Yugoslavia—seemed to be going well for the United States.[157] The article notes, "In China last week, the U.S. pulled out the final sagging props that had held up its policy, and a lot of decayed timbers were exposed in the process. The old structure, never sound, was disintegrating, all right, and it made a portentous heap of rubble when it fell. The tragedy was that the U.S. had nothing to put in its place." The article details *Time's* version of the many failures of American policy in China through World War II and the Chinese civil war, which *Time* describes (with a reference to the Marshall Mission) as an "eerie blend of fatalism and foolishness." Portraits of Stilwell, Marshall, the Chiangs, and Ambassador Stuart head a page devoted to the "platitudes and principles" of America's policy. The article concludes that "at no point in the long chronicle of its failure had [American foreign policy toward China] displayed a modest fraction of the stamina and decisiveness which had checked Communism in Europe." As detailed above, of course, it was Truman's European policy that prompted China supporters in Congress to push for more aid to an ailing Chiang. In any case, the story follows the basic line that America, if it had wanted to, could have decisively influenced events in China for the better, and by inaction, naïveté, and wrongheadedness had instead driven China to Communism.

In the autumn months of 1949, there were few China stories or pictures in *Time* magazine; almost all are devoted to portraits and maps of the Communist absorption of the mainland, city by city, with more discourses on Communist history. The issue of November 21 introduces visually the topic that would be at the center of Sino-American relations for more than two decades. In a cartoon, figures representing the United States (Acheson) and Britain (Churchill) sit at a table marked "U.N. Council" (see figure 1.1).

Between Acheson and Churchill is an empty seat labeled "China." The men are turned to face a dumpy figure in a starred cap and Chinese military uniform labeled "Mao" holding a portfolio labeled "Cold War." Acheson says, "Good heavens—is that Chiang?" Churchill (the cartoon is from a British newspaper) answers, "Even if he isn't, he is practically unrecognizable." The accompanying article, significantly titled "Toward Recognition," sums up the confusion and ambiguity in American intentions. On the one hand, the reader is told, "The U.S. State Department wants to recognize the Chinese Communists," but further on it is noted that the president "has decided, somewhat belatedly, that he doesn't like the Chinese Communists." The "belatedly" is a characteristic *Time* snide remark at the commander in chief. The article also drops hints that a limitation of Soviet influence is possible, since "the presence of U.S. diplomats would tend to 'inhibit' Russian moves to strengthen the Russian grip on China." Again, one could imagine the reaction of Russian diplomats reading such stories

Figure 1.1. November 21, 1949: Cartoon, "Question of Recognition." United States (Acheson) and Britain (Churchill) sit at a table marked "U.N. Council." (Credit: David Low in Evening Standard. *Reprinted by permission from Solo Syndication, London/Associated Newspapers.)*

and assuming, in their Cold War paranoia, that they were based on the substance of Sino-American conversations.

Another significant image in the last weeks of 1949 is a map in the issue of December 12 of the "Western horizons" of the Pacific. At the top is an outline of Asia almost completely dominated by a red swath. Rolling across the Pacific is the fortress of the "horizon of U.S. defense." At the center and the apex of the map is Formosa, topped by a Nationalist flag (see figure 1.2).

In case the reader-viewer feels any confusion about the island's importance, it is surrounded by the bright white edges of a citadel. The accompanying article, "Time for Action?" outlines the growing case for holding Formosa, since it "would demonstrate to all of Asia the determination of the U.S. to stand fast." Notably, South Korea, while not shown to be under the Red zone of Communism, is also *not* placed within the horizon of U.S. defense. Here *Time* engages in question begging. The U.S. government was in the process of trying to find a way to cut the last strings from the island-bound Republic of China; only *Time* contemplated "action."

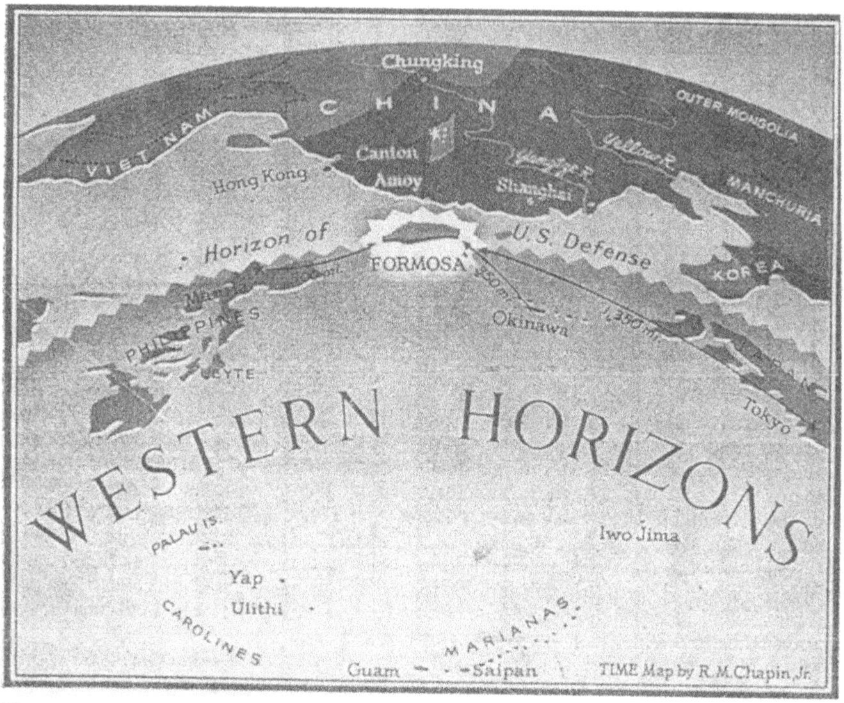

Figure 1.2. December 12, 1949: Map of "Western horizons" of the Pacific. (Artist: R. M. Chapin Jr. Copyright 1949 by Time, Inc. Reprinted with permission.)

Seemingly as a capstone to the progress of history, or *Time*'s view of it, the last issue of the year, dated December 26, shows a picture of "Mao in Moscow" marching with Soviet leaders. The article declares that speculation over China going the route of Tito is now over, and "sworn enemies of the West" control China. As if to lay aside any further ambiguity, the article announces: "Joseph Stalin had much to celebrate" from the "Moscow-Beijing axis." The last picture of China of that year is a full-drawn map of Formosa. Noted are some of the troubles of the Nationalist citadel, with the local population and among themselves (the graphics of the map include ports, airfields, pineapple plantations, and a region inhabited by "headhunters"). The article argues, "The only possible source of . . . help [to hold Formosa] was the United States which, if it wanted to, could deny Formosa to the Communists at little risk to itself." The article ends with a strong argument in favor of this policy option, but doubts whether the Truman administration would pursue it. As we shall see in the next chapter, Chiang and Washington would soon resume their alliance, although both would be surprised at the cause, an undeclared but bloody war between America and China on the Korean Peninsula (a place *Time* did not list or picture as being part of the wall of containment against Communism). With the sound of the guns, China would be shape-shifted again, this time into a beast of menace, razor-sharp in tooth and claw.

Notes

1. Alexis de Tocqueville, *Democracy in America,* trans. G. Lawrence, ed. J. P. Mayer (Garden City, N.Y.: Doubleday, 1969), 91n50.

2. A. O. Aldridge, *The Dragon and the Eagle: The Presence of China in the American Enlightenment* (Detroit: Wayne State University Press, 1993), 8; J. Ching & W. G. Oxtoby, eds., *Discovering China: European Interpretations in the Enlightenment* (Rochester, N.Y.: University of Rochester Press, 1992).

3. J. Gray, *Rebellions and Revolutions: China from the 1800s to the 1980s* (Oxford and New York: Oxford University Press, 1990), 3.

4. Gray, *Rebellions and Revolutions,* 1.

5. D. L. Anderson, *Imperialism and Idealism: American Diplomats in China, 1861–1898* (Bloomington Indiana University Press, 1985), 3.

6. J. K. Fairbank, "Our Inherited China Policy," in *The United States and China,* (Cambridge, Mass.: Harvard University Press, 1983), 307–35.

7. T. H. Hardin, "American Press and Public Opinion in the First Sino-Japanese War," *Journalism Quarterly* 50 (January 1973): 54–59; G. P. Jan, "Public Opinion and American Policy toward the Sino-Japanese War, 1937–1945," *Asian Profile* 7 (April 1979): 319–34.

8. H. R. Isaacs, *Scratches on Our Minds; American Images of China and India* (New York: J. Day, 1958).

9. Aldridge, *The Dragon and the Eagle,* 99.

10. Isaacs, *Scratches on Our Minds,* 71.

11. Sadako Ogata, *Defiance in Manchuria: The Making of Japanese Foreign Policy, 1931–32* (Los Angeles: Center for Japanese and Korean Studies, 1964); T. Yoshihashi, *Conspiracy at Mukden: The Rise of the Japanese Military* (New Haven, Conn.: Yale University Press, 1963).

12. Donald J. Friedman, *The Road from Isolation: The Campaign of the American Committee for Non-Participation in Japanese Aggression, 1938–1941* (Cambridge, Mass.: Harvard University Press, 1968), 1.

13. Michael Schaller, *U.S. Crusade in China: 1938–1945* (New York: Columbia University Press, 1979), 32.

14. Foster Rhea Dulles, *China and America: The Story of Their Relation Since 1784* (Washington, D.C.: Kennikat Press, 1967), 221.

15. Stephen G. Craft, "Peacemakers in China: American Missionaries and the Sino-Japanese War, 1937–1941," *Journal of Church and State* 41 (1999): 575–91.

16. Zhang Baijia, "Kang Ri Zhangzheng Shiqi Guogong Liangdang De Dui Mei Zhengce" [Guomindang and Communist policies toward the United States during the anti-Japanese war], trans. Shannon Wang, in *Zhong Mei Guanxi Shi Shang Chenzhong De*

Yiye [Sino-American Relations, 1944–1955: Collaborative reassessment of a troubled time], eds. Yuan Ming and Harry Harding (Beijing: Beijing University Press, n.d.), 19–20.

17. Han Suying, *Birdless Summer* (New York: Putnam, 1968), 235–36.

18. Zhang Baijia, "Kang Ri Zhanzheng," trans. Shannon Wang, 19–20.

19. Herbert Feis, *The China Tangle: The American Effort in China from Pearl Harbor to the Marshall Mission* (Princeton, N.J.: Princeton University Press, 1953), 4–5.

20. Theodore White, *In Search of History* (New York: Harper & Row, 1978), 130. See also: W. A. Swanberg, *Luce and His Empire* (New York: Charles Scribner's Sons, 1972), 189.

21. Xiaohua Ma, "The Sino-American Alliance During World War II and the Lifting of the Chinese Exclusion Acts," *American Studies International* 38 (2000): 39–61.

22. *New York Times*, December 9, 1941.

23. Jonathan Spence, *To Change China: Western Advisers in China, 1620–1960* (Boston: Little, Brown, 1969), 234.

24. Winston S. Churchill, *The Second World War*, vol. 4: *The Hinge of Fate* (London: Cassell, 1951), 119.

25. Winston Churchill, *Closing the Ring* (Boston: Houghton Mifflin, 1951), 328.

26. Cordell Hull, *The Memoirs of Cordell Hull*, vol. 2 (New York: Macmillan, 1948), 1583.

27. Hull, *The Memoirs*, 1587.

28. United States Department of State, *Foreign Relations of the United States: Diplomatic Papers, 1942, III: Europe* (Washington, D.C.: Government Printing Office), 580–81.

29. A. F. M. Shamsur Rahman, "United States Economic and Military Assistance Policy toward China During World War II and Its Immediate Aftermath" (unpublished Ph.D. diss., University of Kansas 1988), 230.

30. Rahman, "United States Economic and Military Assistance," 3.

31. Rahman, "United States Economic and Military Assistance," 231.

32. Barbara Tuchman, *Stilwell and the American Experience in China, 1911–45* (New York: Macmillan, 1971), 250.

33. For overview, see: Christopher Thorne, *Allies of A Kind: The United States, Britain, and the War against Japan* (Oxford: Oxford University Press, 1978); Xiaoyuan Liu, *A Partnership for Disorder: China, the United States, and Their Policies for the Postwar Disposition of the Japanese Empire, 1941–1945* (Cambridge: Cambridge University Press, 1996).

34. Thomas Doherty, *Projections of War: Hollywood, American Culture and World War II* (New York: Columbia University Press, 1993), 137.

35. Isaacs, *Scratches on Our Minds*, 167.

36. As quoted in Clayton R. Koppes and Gregory D. Black, *Hollywood Goes to War: How Politics, Profits and Propaganda Shaped World War II Movies* (Berkeley: University of California Press, 1987), 236.

37. Frank Capra, *The Name above the Title: An Autobiography* (New York: Macmillan, 1971).

38. Erik Barnouw, *Documentary: A History of the Non-Fiction Film* (New York: Oxford University Press, 1983), 155–62.

39. *New York Times*, November 21, 1943.

40. Graham Peck, *Two Kinds of Time* (Boston: Houghton Mifflin, 1950), 35.

41. Nathaniel Peffer, "Our Distorted View of China," *New York Times Magazine*, November 7, 1943.

42. Kenneth E. Shewmaker, *Americans and Chinese Communists, 1927–1945* (Ithaca, N.Y.: Cornell University Press, 1971); David L. Wilson, "The Attitudes of American Consular and Foreign Service Officers toward Bolshevism in China, 1920–1927" (Ph.D. diss., University of Tennessee 1974).

43. Warren I. Cohen, *America's Response to China: A History of Sino-American Relations* (New York: Columbia University Press, 1990), 136–39.

44. Schaller, *U.S. Crusade in China*, 190.

45. Yawei Liu, "The United States According to Mao Zedong: Chinese-American Relations, 1893–1976" (Ph.D. diss., Emory University 1996), 72.

46. Liu, "United States According to Mao Zedong," 73.

47. Liu, "United States According to Mao Zedong," 73.

48. Liu, "United States According to Mao Zedong," 75.

49. Liu, "United States According to Mao Zedong," 76.

50. Liu, "United States According to Mao Zedong," 114.

51. Schaller, *U.S. Crusade in China*, 20.

52. Gary Dean Best, *Pride, Prejudice, and Politics: Roosevelt versus Recovery, 1933–1938.* (Westport, Conn.: Praeger, 1993), 3.

53. Charles Romanus and Riley Sunderland, *Stilwell's Mission to China* (Washington, D.C.: Department of the Army, 1953), 121.

54. Schaller, *U.S. Crusade in China*, 190.

55. Schaller, *U.S. Crusade in China*, 178.

56. Herbert Feis, *The China Tangle: The American Effort in China from Pearl Harbor to the Marshall Mission* (Princeton, N.Y.: Princeton University Press, 1953), 21.

57. United States Department of State, *U.S. Relations with China* (Washington, D.C.: Government Printing Office), 473–76.

58. United States Department of State, *U.S. Relations with China*, 510.

59. For a laudatory account see: J. Israel, "Mao's Mr. America: Edgar Snow's Images of China," *Pacific Historical Review* 47, no. 1 (1978): 107–22. For an overview of Snow's strengths and weaknesses as an observer of China, see: J. M. Hamilton, *Edgar Snow: A Biography* (Bloomington: Indiana University Press, 1988).

60. Hamilton, *Edgar Snow*, 77.

61. Cited in Hamilton, *Edgar Snow*, 77.

62. Alain Jaubert, *Making People Disappear: An Amazing Chronicle of Photographic Deception* (Washington, D.C.: Pergamon-Brassey's, 1989[1986]), 104.

63. Theodore H. White, *In Search of History: A Personal Adventure* (New York: Harper & Row, 1978), 245–49.

64. White, *In Search of History*, 254.

65. Thomas Griffith, *Harry and Teddy: The Turbulent Friendship of Press Lord Henry R. Luce and His Favorite Reporter, Theodore H. White* (New York: Random House, 1995), 270.

66. Jonathan Fenby, *Chiang Kai-Shek: China's Generalissimo and the Nation He Lost* (New York: Carroll & Graf, 2004), 345.

67. Guangqiu Xu, "The Issue of U.S. Air Support for China During the Second World War, 1942–45," *Journal of Contemporary History* (Great Britain) 36, no. 3 (2001): 459–84.

68. Yufeng Wang, "John S. Service and United States-China Relations, 1933–1945" (Ph.D. diss., University of West Virginia, 1997), 48.

69. Wang, "Service and United States-China Relations," 48.

70. Charles F. Romanus and Riley Sunderland, *United States Army in World War II: China-Burma-India Theater: Stilwell's Command Problems* (Washington, D.C.: Office of the Chief of Military History, Department of the Army, 1956), 302.

71. See chronology of campaign in: Barbara W. Tuchman, *Stillwell and the American Experience in China, 1911–45* (New York: Bantam, 1972); Fred Eldridge, *Wrath in Burma: The Uncensored Story of General Stilwell and International Maneuvers in the Far East* (New York: Doubleday, 1946); Charles F. Romanus and Riley Sunderland, *Time Runs Out in CBI* (Washington, D.C.: Office of the Chief of Military History, Department of the Army, 2003); Chang Hsien-Wen, ed., *K'ang-Jih Chan-Cheng Ti Cheng -Mien Chan-Ch'ang 1959* trans. Shannon Wang (n.p.: Honan Jen-Min Ch'u-Pan-She, 1987).

72. Herbert Feis, *The China Tangle: The American Effort in China from Pearl Harbor to the Marshall Mission* (Princeton, N.J.: Princeton University Press, 1953), 43.

73. Feis, *The China Tangle*, 63.

74. Hsi-Sheng Ch'I, "The Military Dimension, 1942–1945," in *China's Bitter Victory: The War With Japan, 1937–1945*, eds. James C. Hsiung and Steven I. Levine (Armonk, N.Y.: M. E. Sharpe, 1992), 157–84.

75. Joseph W. Stilwell, The Stilwell Papers. arranged and ed. Theodore H. White (New York: William Sloane Associates, 1948), 316–19.

76. The following are characteristic notes Stilwell wrote to himself in preparing for meetings with the President: "FDR. Recommendations. Private Army of One Corps. Keep X-Force and add one in China. Recommend to Peanut more power for me. Field Chief of Staff. Oust running dog. FDR. My mission complicated by not knowing what direct messages [from Roosevelt to Chiang] contain. No bargaining power." Quoted in Romanus and Sunderland, *Stilwell's Command Problems*, 63.

77. Feis, *The China Tangle*, 18.

78. Romanus and Sunderland, *Stilwell's Command Problems*, 383–84.

79. Don Lohbeck, *Patrick J. Hurley* (Chicago: Henry Regnery, 1956), 294–96.

80. Yawei Liu, "United States According to Mao Zedong," 119.

81. Yawei Liu, "United States According to Mao Zedong," 118.

82. All Service material is drawn from: Yufeng Wang, "John S. Service and United States-China Relations, 1933–1945," (Ph.D. diss., University of West Virginia, 1997); material below on the Dixie Mission is drawn from: Yawei Liu, "The United States According to Mao Zedong" and David D. Barrett, *Dixie Mission: The United States Army Observer Group in Yanan, 1944* (Berkeley: Center for Chinese Studies, 1970); also consulted was: Anne-Marie Brady, *Making the Foreign Serve China: Managing Foreigners in the People's Republic* (Lanham, Md.: Rowman & Littlefield, 2003).

83. Robert A. Diamond, ed., *China and U.S. Foreign Policy* (Washington, D.C.: Congressional Quarterly, 1971), 1.

84. Robert M. Blum, *Drawing the Line: The Origins of the American Containment Policy in East Asia* (New York: W. W. Norton, 1982); Nancy B. Tucker, *Patterns in the Dust: Chinese-American Relations and Recognition Controversy, 1949–1950* (New York: Columbia University Press, 1983).

85. Yawei Liu, "United States According to Mao Zedong," 125.

86. Barrett, *Dixie Mission,* 74.

87. United States Department of State, *Foreign Relations of the United States: Diplomatic Papers, 1944,* vol. 6, *China* (Washington, D.C.: Government Printing Office), 727–29.

88. United States Department of State, *Foreign Relations of the United States: Diplomatic Papers, 1945,* vol. 7, *The Far East: China* (Washington, D.C.: Government Printing Office), 318–19.

89. United States Department of State, *1945,* vol. 7, *The Far East: China,* 272–78.

90. United States Department of State, *1945,* vol. 7, *The Far East: China,* 310–17.

91. United States Department of State, *1945,* vol. 7, *The Far East: China,* 317–22.

92. Yawei Liu, "United States According to Mao Zedong," 132.

93. Tsou Tang, *America's Failure in China, 1941–1950* (Chicago: University of Chicago Press, 1963), 541.

94. Yufeng Wang, "John S. Service and United States-China Relations, 1933–1945" (Ph.D. diss., University of West Virginia, 1997).

95. Joseph W. Esherick, ed., *Lost Chance in China: The World War II Despatches of John S. Service* (New York: Random House, 1974), xix.

96. E. J. Kahn, Jr., *The China Hands: America's Foreign Service Officers and What Befell Them* (New York: Viking Press, 1975), 192.

97. Yafeng Xia, "From Antagonists to Adversaries: United States-China Talks During the Cold War, 1949–1972" (Ph.D. diss., University of Maryland, 2003), 23.

98. David Mayers, *George Kennan and the Dilemmas of U.S. Foreign Policy* (New York: Oxford University Press, 1988), 177.

99. Thomas J. Christensen, "Domestic Mobilization and International Conflict," (Ph.D. diss., Columbia University, 1993),181.

100. Jonathan Fenby, *Chiang Kai-Shek: China's Generalissimo and the Nation He Lost* (New York: Carroll & Graf, 2004), 475.

101. Fenby, *Chiang Kai-Shek,* 475.

102. Joseph Alsop, "The Feud between Stilwell and Chiang," *Saturday Evening Post,* January 7, 1950, 16.

103. Tang Tsou, *America's Failure in China, 1941–50* (Chicago: University of Chicago Press, 1963), 218.

104. W. I. Cohen, "The China Lobby," in *Encyclopedia of American Foreign Policy: Studies of the Principal Movements and Ideas,* ed. A. Deconde (New York: Scribner, 1978), 104–10.

105. W. I. Cohen, "The China Lobby," 104.

106. John Tebbel, *The Life and Good Times of William Randolph Hearst* (New York: Dutton, 1970), 327–43; Ferdinand Lundberg, *Imperial Hearst: A Social Biography* (Westport, Conn.: Greenwood, 1953), 343–47; Warren Cohen, *The Chinese Connection* (New York: Columbia University Press, 1978), 260, 287–89; Ross Koen, *The China Lobby in*

American Politics (New York: Macmillan, 1960), 27–53; Joseph C. Keeley, *The China Lobby Man: The Story of Alfred Kohlberg* (New Rochelle, N.Y.: Arlington House, 1969), 114; Tucker, *Patterns in the Dust*, 81–99.

107. Arthur Vandenberg, Jr., and Arthur H. Sense, *The Private Papers of Senator Vandenberg, 1944–1951* (Boston: Houghton Mifflin, 1952), 532.

108. See John S. Service, *The Amerasia Papers: Some Problems in the History of U.S.-China Relations* (Berkeley: University of California Press, 1971). Also see: Barrett, *Dixie Mission*; Dorothy Borg, ed., *Uncertain Years: Chinese-American Relations, 1947–1950* (New York: Columbia University Press, 1970); Tucker, *Patterns in the Dust.*

109. Service, *The Amerasia Papers*, 191–92.

110. The term comes from a 1850s treatise by scholar Wei Yüan, "Treatise on Maritime Kingdoms" (Haikuo T'u-Chih) adapted from a book by Lin Tse-Hsu; see Michael H. Hunt, *The Making of a Special Relationship: The United States and China to 1914* (New York: Columbia University Press, 1983), 45.

111. Harvey Kehr, *The Amerasia Spy Case: Prelude to McCarthyism* (Chapel Hill: University of North Carolina Press, 1996).

112. United States Department of State, Diplomatic Papers, vol. 7, *The Far East: China*, 88–103.

113. Chen Jian, *China's Road To the Korean War: The Making of the Sino-American Confrontation* (New York: Columbia University Press, 1990), 246.

114. Enlai Zhou, *Report on Problems Concerning Peace Talks. Selected Works of Zhou Enlai, vol. 50* (Beijing: Foreign Language Press, 1981), 360–64; Zedong Mao, *On the People's Democratic Dictatorship. Selected Works of Mao Tsetung, vol. 4* (Bejing: Foreign Language Press, 1967), 416.

115. William W. Stueck, Jr. *The Road to Confrontation: American Policy toward China and Korea, 1947–1950* (Chapel Hill: University of North Carolina Press, 1981), 121–22.

116. Warren I. Cohen, "Acheson, His Advisers, and China. 1949–1950," in *Uncertain Years: Chinese-American Relations, 1947–1950*, ed. Dorothy Borg (New York: Columbia University Press, 1980), 49–51.

117. Tucker, *Patterns in the Dust*, 176, 188.

118. Warren Cohen, "Research Note: Conversations with Chinese Friends: Zhou Enlai's Associates Reflect on Chinese-American Relations in the 1940's and the Korean War," *Diplomatic History* 11, no. 3 (1987): 285.

119. Warren Cohen, "Conversations with Chinese Friends."

120. Zhai Qiang, "Recent Chinese Writings on 1945–1955 Sino-U.S. Relations," *SHAFR Newsletter*, 20(4) (1989): 75–84.

121. Zhai Qiang, "Recent Chinese Writings"

122. K. S. Karol, *The Second Chinese Revolution*, trans. Mervyn Jones (New York: Hill and Wang, 1973), 48.

123. Chen Jian, *China's Road to the Korean War*, 40.

124. Chen Jian, *China's Road to the Korean War.*

125. Yuan Ming, "The Failure of Perception: America's China Policy, 1949–1950," in *Sino-American Relations, 1945-1955: A Joint Reassessment of a Critical Decade*, eds. Harry Harding and Yuan Ming (Wilmington, Del.: Scholarly Resources, 1989), 151.

126. United States Department of State, *Foreign Relations of the United States: Diplomatic Papers, 1949.* vol. 9, *The Far East: China* (Washington, D.C.: Government Printing Office), 121–22.

127. United States Department of State, *Diplomatic Papers, 1949,* vol. 9, *The Far East: China* (Washington, D.C.: Government Printing Office), 115.

128. United States Department of State, *Diplomatic Papers, 1949,* vol. 9, *The Far East: China* (Washington, D.C.: Government Printing Office), 243–44.

129. United States Department of State, *Diplomatic Papers, 1950,* vol. 6, *East Asia and the Pacific* (Washington, D.C.: Government Printing Office), 289.

130. *China and U.S. Far East, 1945–1967* (Washington, D.C.: Congressional Quarterly Service, 1967), 48.

131. Numbers and points in this paragraph are drawn from: Chen Jian, *China's Road to the Korean War: The Making of the Sino-American Confrontation* (New York: Columbia University Press, 1990), 98–101.

132. Allen S. Whiting, "Mao, China, and the Cold War," in Akira Iriye and Yonosuke Nagai, *The Origins of the Cold War in Asia* (New York: Columbia University Press, 1977), 253.

133. Jian, *China's Road to the Korean War,* 42.

134. Michael W. Sheng, *Battling Western Imperialism: Mao, Stalin, and the United States* (Princeton, N.J.: Princeton University Press, 1997), 195.

135. Sheng, *Battling Western Imperialism,* 163.

136. John W. Garver, "Polemics, Paradigms, Responsibility, and the Origins of the U.S.-PRC Confrontation in the 1950s," *Journal of American-East Asian Relations* 3, no. 1 (1994): 1–34.

137. Gene David McLean, "American Nationalism, the China Myth, and the Truman Doctrine: The Question of Accommodation with Peking, 1949–1950," *Diplomatic History* 10, no. 1 (1986): 30.

138. Michael Schaller, "Consul General O. Edmund Clubb, John P. Davies, and the Inevitability of Conflict between the United States and China, 1949–1950: A Comment and New Documentation," *Diplomatic History* 9, no. 2 (1985): 149, 153.

139. David A. Mayers, *Cracking the Monolith: U.S. Policy Against the Sino-Soviet Alliance, 1949–1955* (Baton Rouge: Louisiana State University Press, 1986), 30–31.

140. George Kennan, quoted in John Lewis Gaddis, *Strategies of Containment: A Critical Appraisal of Postwar American National Security Policy* (New York: Oxford University Press, 1982), 41.

141. David A. Mayers, *Cracking the Monolith,* 24.

142. Jian Chen, "China's Road to the Korean War," 31.

143. Tang Tsou, *America's Failure in China, 1941–50* (Chicago: University of Chicago Press, 1963): 475.

144. Christensen, "Domestic Mobilization and International Conflict," 190.

145. Christensen, "Domestic Mobilization and International Conflict," 194.

146. Robert Blum, *Drawing the Line,* 8.

147. NSC 6, "The Position of the U.S. Regarding Short-Term Assistance to China," March 26, 1948. Foreign Relations of the United States. 8: The Far East (China), 44–50.

148. Mayers, *Cracking the Monolith,* 19–21.

149. Dean Acheson, *Present at the Creation: My Years in the State Department* (New York: W. W. Norton, 1969), 306.

150. Foster Rhea Dulles, *American Policy toward Communist China* (New York: Thomas Y. Crowell, 1979), 49–51.

151. Adam Ulam, *The Rivals: America and Russia Since World War II* (New York: Penguin, 1980), 138.

152. United States Department of State, *Foreign Relations of the United States: Diplomatic Papers, 1948*, vol. 7, *The Far East: China* (Washington, D.C.: Government Printing Office), 707–18; United States Department of State. *Foreign Relations of the United States: Diplomatic Papers, 1949*. vol. 8, *The Far East: China* (Washington, D.C.: Government Printing Office), 16–19.

153. Mayers, *Cracking the Monolith*, 175.

154. Yafeng Xia, "From Antagonists to Adversaries: United States-China Talks During the Cold War, 1949–1972" (Ph.D. diss., University of Maryland, 2003), 27.

155. United States Department of State, *Foreign Relations of the United States: Diplomatic Papers, 1948*, vol. 8, *The Far East: China* (Washington, D.C.: Government Printing Office), 146–55.

156. David D. Perlmutter, "Visual Images and Foreign Policy: Picturing China in the American Press, 1949–1989" (Ph.D. diss., Minneapolis: University of Minnesota Press, 1996).

157. Adam Ulam, *The Rivals: America and Russia Since World War II* (New York: Penguin, 1980), 138.

CHAPTER 2

The Korean War, 1950–1953

> "[T]his Chinese Communist advance into North Korea . . . was one of the most terrific disasters that has occurred to American foreign policy, and certainly . . . the greatest disaster which occurred to the Truman Administration. It did more to destroy that administration, and did more to destroy and undermine American foreign policy than anything that we know about—the whole Communists in government business, the whole corruption outcry, was really just window-dressing put upon this great disaster."
>
> Secretary of State Dean Acheson[1]

1. Introduction: Why Korea?

If the weight of evidence of modern scholarship on Sino-American relations holds that the "lost chance" never existed, even its proponents think any opportunity for rapprochement between the two antagonists had faded considerably by early 1950. On February 14, Mao, acting on his decision to "lean to one side," signed a treaty of alliance with the Soviet Union. But many in Washington did not think that the door to China was completely shut, and probably would have been amazed if told that within the year Chinese and American armies would be clashing, and further, that the countries would be without formal relations for decades to come. Hope seemed to spring, as of December 1949, from the fact that there was a split in the American government on the one issue that would be the most divisive. Secretary of State Dean Acheson was ready to write off Formosa as an impediment to negotiations with the Chinese.[2] In January 1950, a State Department policy paper was leaked that specifically foretold

Formosa's imminent fall.[3] In opposition, the then secretary of defense Louis Johnson, the Joint Chiefs of Staff, and General Douglas MacArthur formed a cabal within the government to lobby for continued military aid to Chiang.[4] Truman seemed to favor the diplomats. In a speech on January 5, 1950, which, a few months later, he probably wished he had never made, he claimed that the United States had no interests in the island, and that it was, by international law, an integral part of a sovereign China: "The United States has no desire to obtain special rights or privileges or to establish military bases in Formosa at this time. Nor does it have any intention of utilizing its armed forces to interfere in the present situation. The United States government will not pursue a course which will lead to involvement in the civil conflict in China. Similarly, the U.S. government will not provide military aid or advice to Formosa."[5]

As a follow-up, Acheson and the State Department inaugurated a publicity effort arguing that Chiang was doomed, Formosa was part of China, and America had to recognize the new order in the Far East.[6] Among the reasons to abandon the Republic of China was that in the last stage of the civil war any interference was unwise because the United States would "not wish to create a Formosa . . . issue about which the Chinese Communists could rally support within China."[7] In March the momentum in the United Nations also ran toward recognition of the PRC when the secretary-general made a statement that seating at the international body should be based on the political power within a country that had "effective authority" and was "habitually obeyed by the bulk of its population."[8] The State Department also pointed out that any further military aid to Taiwan would just be confiscated by the Red Chinese when (not if) they retook the island. It was thus the firm position of the U.S. government by June 1950 that it would not stop the Communists from invading Formosa and, in effect, would stand by to "let the dust settle" in East Asia.[9] There is even evidence that one faction of American diplomats and intelligence personnel planned some sort of an anti-Chiang coup sometime in the summer.[10]

The China issue had become too symbolic in American politics for a "dump Chiang" campaign to succeed quickly,[11] and in any case the CCP did not interpret America's laissez-faire position toward Formosa as a positive stance for themselves.[12] In China, such "peace moves" by the American diplomats were ignored, ridiculed, or seen as evidence of further imperialist deviousness. In "leaning to one side," Mao had only partly allayed Stalin's suspicions that China could become another Yugoslavia. A memo from a January 8 Chinese Politburo meeting stated that the United States was trying to "sabotage" the revolution.[13] Such views revealed a characteristic dynamic of Mao in particular and Communist leadership in general. In their theoretical system, America was an enemy, and *anything* America did was proof of this fact, whatever America's actual intentions. Moreover, as noted, Mao was more likely to be aggressive when he felt that he was boosting revolutionary "morale" while undermining that of the enemy.[14]

In Mao's view, then, domestic revolution and world revolution were intertwined. As another historian, Qiang Zhai, has put it, "[W]hen Mao committed himself to revolution, he was determined to transform not only China but also the world."[15] In turn, America's perception that it was reaching out to China in conciliation would contribute to its own sense of aggrievement when the Korean War erupted in June 1950. Americans would then perceive the PRC as a Soviet subordinate ally in position to be a "springboard for a further incursion" in Asia.[16] In addition, in the eyes of U.S. policymakers, the press, and the public, Formosa would metamorphose from a sad, doomed remnant of a fallen regime to a stronghold of the global war against Communism, and would appear to be a possible staging point for the defeat of "Red" China. Indeed, the salvation of Formosa would almost lead America to nuclear war twice in the middle and late 1950s.

I will argue here that misguided visions of China almost certainly deceived us into disaster in Korea. First, the invasion of the South by the North Korean People's Army stunned the world. Then, in the late winter months of the year, after a U.S. and United Nations army had beaten back the invaders and driven north toward the Chinese border to complete the unification of Korea, America was surprised again by a massive Chinese intervention. Each "surprise" was, as detailed below, a case of consensual myopia among U.S. policy, military, and press elites, although for some the myopia may have been purposive. And it was dutifully cheered on by Henry Luce's *Time* magazine. The publication noted on October 6, 1950—while three hundred thousand Chinese soldiers were pouring into North Korea—that "Communist premier Zhou Enlai's threat that China 'will not stand aside should the imperialists wantonly invade' North Korea was only propaganda. If Russia or China intended to intervene in Korea, they should have done so earlier when they could have pushed U.N. forces into the sea." Such Luce-think was expressible only in serious discourse because the highest generals and diplomats looked toward the East and saw what they wanted to see: China, although evil and a puppet of the Soviet Union, was impotent in its fury. Instead, we know today that the People's Republic of China had its own agenda, that Moscow was both the manipulator and the frustrater—but not the commander—of Chinese ambitions, and that the Communist Chinese leadership from the earliest days of the war had every intention of throwing American troops off the peninsula.

Korea's tragedy is often overshadowed by other American foreign wars, but the human cost of the 1,100-day "police action," as Truman called it, was great despite the restriction of its battles to an area no larger than New York and Connecticut. At least one million South Koreans died from 1950 to 1953. Of the ten million estimated to have made up North Korea's population in 1950, about a quarter died during the war, from various causes including starvation, disease, bombing, and combat.[17] The death toll of Americans due to battle-related injuries alone was nearly thirty-four thousand, and at least three thousand soldiers from other UN contingents also died.[18] As many as a million Chinese soldiers perished as well, but even the extremely favorable ratio of friendly-to-enemy dead led no American politician to claim that Korea was a World War II-type triumph. Consider this description of the survivors of the 1st Marine Division's epic "March to the Sea" from the Battle of the Chosin Reservoir after being surrounded by hugely superior Chinese forces:

> Every exposed foot of fabric, it seemed, contained its own telltale marks: streaks of syrup from half-frozen C-ration fruit cans; blotches of carbon and hair tonic from cleaning heavily used weapons in subzero temperatures; unmistakable evidence of rampant dysentery; mucous stains from chronic runny noses; and the universal black and scarlet patches of dried human blood—the Marine's, his buddy's, the enemy's.[19]

And these men were the *victors*, shredding whole regiments—up to twenty-five thousand killed—of "Chinese volunteers" in their fighting retreat. In the greater picture, however, the Korean conflict, even though it was classified as a "limited" war, marked the only time that American armies had been defeated on the battlefield, retreated, and then never regained the lost ground.[20] The war's events and outcomes would also poison Sino-American relations for twenty years. Until President Richard Nixon's China visit in 1972, a strong majority of the American public and elites would oppose even the prospect of having full relations with mainland China or allowing it a seat in the United Nations. A final offshoot of the Korea conflict was a war economy: the United States military budget, by May 31, 1951, stood at $48 billion, four times the figure before the war.[21]

How could Americans have been so "surprised" in 1950 by China's intentions and actions? The news stories and pictures in *Time* of 1950 to 1953 offer an insight. With severed ties and outright military conflict during those years, a curious phenomenon affected what Americans saw in newsmagazines about China: since foreign reporters were banned from the PRC (by its government) or banned from traveling to the PRC (by Western governments), most of the actual photos of "China" originated from PRC government media such as the Eastfoto news service. These images were propaganda stills, fakes, or shots of pseudo-events. The student of any photographic corpus from a totalitarian country will be familiar instantly with all the genres of smiling leaders, cheering crowds, valiant soldiers marching, children holding posters of the beloved leader, and so on. The choice for the U.S. newsmagazines, however, was between having no pictures from China and using pictures from China that showed the opposite of what was the editorial stance of the publications. The solution, documented below in the pictorial analysis section of this chapter, was simple and devastating and would become the standard for several decades. In almost all cases a technique counter to the rules of photojournalism is employed: the captions attack the images they describe. The "Red" visual propaganda is mocked, questioned, and discounted by contrarian words. That this is itself an act of (counter)propaganda is, of course, not mentioned or acknowledged. In sum, China was the enemy, a pawn of the Soviets, and so information from China was assumed to be lies. It made sense that when China indicated to America that it planned to intervene in Korea, that message too was discounted.

2. Invasion and Inchon

On Sunday, June 25, 1950, at about 4 a.m. (Korean Standard Time), the armed forces of North Korea (the Democratic People's Republic of Korea Army, or DPRK) began a bombardment of forces of South Korea (the Republic of Korea, or ROK) across the 38th parallel, the line of demarcation between the two sections of Korea established at the end of World War II.[22] DPRK radio announced that the attack was in response to Southern incursions. (This was a lie—the invasion had been planned for months.) By noon, the North declared war, and its land and air forces crossed the border. As the Southern army and the tiny U.S. troop presence reeled back, the shock wave of the first "hot" engagement of the Cold War traveled across the world. U.S. president Harry S. Truman quickly returned to Washington from his home in Missouri. At the same time, the UN Security Council passed a resolution condemning the North Korean attack and called for the immediate cessation of hostilities and the return of North Korean forces to above the 38th parallel. The nine-to-zero vote (with Yugoslavia abstaining) was possible only because the Soviet delegation was absent—in protest against the refusal to admit the PRC to the world body. North Korea dismissed the vote; the attack continued. Notably, the Soviet delegation stayed absent right through the later votes that declared the UN commitment to South Korea. At the time this was represented as folly, but we know now that Stalin was playing a double game. He did not mind conflict between China and the United States, but he did not want to activate the Sino-Soviet treaty that specified intervention if another power declared war on China.[23]

Over the next day the U.S. administration deliberated its options.[24] Truman was sensitive that going to war, even if not in actual name, required a national will to fight. On the morning of June 26, he noted that the nation's leading dailies, the *New York Times* and the *Washington Post*, had both editorialized in favor of a strong response to the invasion.[25] On Capitol Hill the voices calling for toughness (in terms of aid, not sending more U.S. troops) were Republican critics of the administration, while Democrats adopted a wait-and-see attitude toward the president's actions.[26] In short, no strong antiwar voices or lobby appeared. The press and political, international, and public consensus looked to the White House for leadership. In this light the reasons for a military response seemed irrefutable. South Korea was an American-sponsored creation, with almost its entire budget representing U.S. aid.[27] American thinking in early 1950 was shifting from the "Europe first" (or Europe only) direction in which it had drifted in previous years toward a global strategy of containment of Communism.[28]

That the invasion was a Soviet gambit was not in question in June 1950. State Department intelligence officers reported that North Korea was a Soviet client.[29] The CIA reported a few days after the invasion that it was the product of "Soviet direction and Soviet material support" and was aimed at "the last remaining anti-Communist bridgehead on the mainland of northern Asia, thereby undermining the position of the U.S. and the Western Powers throughout the Far East."[30] Accordingly, Korea was seen through the lens of a global war against Communism in which every battle was a test of the greater struggle, a view that had also been propounded earlier by the National Security Council.[31] On a personal level, Truman was genuinely angered and outraged at what he felt was the perfidious Communists' gross violation of the laws of nations. He linked the Korean invasion to earlier aggressions by totalitarian powers such as the Japanese empire in Manchuria and Hitler at Munich. He observed in his memoirs "how each time that the democracies failed to act it had encouraged the aggressors to keep going ahead. Now Communism was acting in Korea just as Hitler, Mussolini, and the Japanese had acted ten, fifteen, and twenty years earlier."[32]

As a result, June 26 and 27 were feverish days of meetings. A State Department team worked almost around the clock and the next morning drafted several resolutions for Truman.[33] After some rewrites, at 11:30 Truman and most of his advisers met with some congressional representatives, including leading China bloc proponents like California's William K. Knowland. Truman put forth the case that the Security Council resolution empowered the United States to act militarily in Korea to restore the border to the 38th parallel (but he made no mention at the time of the possibility of unifying all of Korea).[34] Truman "characterized the United States combat operations in Korea as a police action for the United Nations against an unlawful bandit attack on the South Korean Republic."[35] At noon a statement was read to the press announcing that the United States would act militarily to stop South Korea from falling to the Communists. The following comment contained in the announcement would become the second major news item of the day:

> The attack upon Korea makes it plain beyond all doubt that Communism has passed beyond the use of subversion to conquer independent nations and war. It has defied the orders of the Security Council of the United Nations issued to preserve international peace

> and security. In these circumstances the occupation of Formosa by Communist forces would be a direct threat to the security of the Pacific area and to the United States forces performing their lawful and necessary functions in that area.[36]

Meanwhile, at the prompting of the United States, the Security Council of the United Nations passed twin resolutions condemning the North Korean invasion and asking for punitive sanctions against North Korea. America was at war (although not officially), and the Korean venture was immediately linked to the last front of the Chinese civil war, the island of Formosa.

We know that Truman's Formosa initiative startled the PRC's leadership, but from an American view the two small quasi countries (Formosa and South Korea) were viewed as fronts in the bigger war. The main impetus for an active alliance with Formosa apparently came from the Joint Chiefs and the Defense Department, acting on an earlier proposal by MacArthur, who had just been appointed commander of UN military forces in Korea. They were concerned that a Communist attack on the island would threaten American forces in Korea or would force the United States to shift air and ground resources from the coming battles on the peninsula.[37] Indeed, Truman expressed the realistic fear that the Korean invasion would be followed up by an attack on Formosa.[38] This fear was, history now shows, unwarranted: the Korea invasion stopped a planned PRC invasion of Taiwan. But the Chinese troop movements away from the coastal areas facing inland and toward Korea were not known at the time.[39]

In the harsh glare of retrospect, Korea's becoming the flash point between East and West was one of those "surprises" of war that, like Pearl Harbor, seemed so easy to predict that some would later assume that America *wanted* to provoke an invasion so that it could respond.[40] Actually, there is no evidence that any factor besides inattention, selective perception, and wishful thinking kept American policymakers from anticipating—and forestalling—the initial Northern invasion of the South. (The same factors were the main inhibitors of accurate prophecy when, four months later, American and UN forces were surprised by a Chinese intervention in the war.) In early 1950, Korea did not seem to anyone to be a lightning rod of world conflict. The ancient country, occupied by Japan since 1910, had been divided at the 38th parallel into Soviet and American zones after its 1945 liberation. The United States and the Soviet Union had agreed that the division would be a temporary stage before elections and national unification. As in the case of Germany, the Soviet Union, in the darkening atmosphere of the Cold War, resisted ballot box determinations, so gradually North Korea and South Korea became bellicose adversaries.

The leader of the North was the head of its Workers' Communist Party, Kim Il Sung. He had fought for years as an anti-Japanese partisan but, at the same time, he had served in the Russian army within a Korean unit. He had, however, no experience with commandership beyond leading a few hundred guerrillas and a small cadre of party loyalists. He dreamed of the reunification of Korea but hoped to accomplish it without the assistance of any foreign troops. His first hope had been that, prodded by Communist guerrillas under his command influence, the South would simply collapse. By March 1949 this was the strategy Stalin had approved, but by the end of the year Kim assessed that only war would bring him the great prize. His rival in the South, President Syngman Rhee, was a hard-line anti-Communist and also wished to reunify the nation under his leadership. Many in Washington considered Rhee to be unstable and were worried that he would attack the North rather than the reverse, so the South Korean army was starved of heavy equipment, especially modern planes, tanks, and anti-tank artillery. Politically, the United States and the UN recognized Rhee's government as the sole government of Korea.[41]

By early 1950, the situation on the peninsula was in a state of tension but not, to American eyes, on the verge of war. Yet Kim's cohort of war commanders were all dedicated to following the Chinese model of reunification through military victory.[42] In Northern newspapers and radio broadcasts, "unification" was the regular bombast. Pro-Communist guerrillas in the South harried the ROK troops and police—although during the war they would prove to be a minor force, not the huge revolutionary army that Kim had promised to Stalin and Mao. As of June, the Northern armies totaled about 135,000 soldiers and militia in all, many concentrated in tough "assault" divisions. Most important, by wartime, Russia had supplied the North with hundreds of World War II vintage T-34 tanks.

The rhetoric from anti-Communist Rhee was a mirror image of the North's. On March 1, 1950, he announced, "We shall respond to the cries of our brothers in distress, even though some of our friends across the sea tell us that we must not cherish thoughts of attacking the foreign puppet who stifles the liberties of our people in the North."[43] Such threats, as journalist and historian Bradley Martin has assessed, "amounted to little more than bravado."[44] The U.S. Army's history of the conflict tersely summed up the state of the ROK army: "The[y] had no tanks, no medium artillery, no 4.2-in. mortars, no recoilless rifles, and no fighter aircraft or bombers."[45] Indeed, while U.S. advisers assisted in training the Southern army, the U.S. troop presence on the peninsula was small and was even drawn down during the period before the Korean War.

Another reason America did not plan for a Korean war was that it was assumed that the next war would be a global confrontation with the Soviet Union and its allies. The atomic bomb—the "war-ender"—seemed in the late 1940s to be the arbiter of all wars to come for the United States.[46] One 1950s critic wrote that "military thinking" was, at first, "paralyzed by [the bomb's] magnitude."[47] In response, all branches of the service prepared for (and invested their shrinking budgets in) the atomic battlefield, which would be dominated by airpower, not infantry platoons or tank brigades. David T. Fautua describes his view:

> The Defense Department claimed that strategic air power alone was the best means. The Joint Chiefs of Staff agreed that strategic strikes should have first priority to deter Soviet expansionism overseas as well as thwart any notions of bomber attacks against the North American continent. The combination of atomic weapons and America's post–World War II global prominence had thus elevated the concept of strategic air power into a strategy of deterrence.[48]

But there was an alternative view, as expressed in National Security Council Paper 68 (NSC 68.1), that America should have a larger, balanced army, capable of responding to a variety of threats and war scenarios; otherwise, the United States and its allies, if involved in smaller nonglobal confrontations that "d[id] not now have the forces in being and readily available to defeat local Soviet moves with local action," would have to "accept reverses or make these local moves the occasion for war—for which we are not prepared."[49]

Worse, the long peace after an exhausting war also contributed to a general sloppiness and lack of training in the armed services. Basic equipment was not replaced, or was poorly maintained and not upgraded. Units were under strength. The low rate and quality of training—even among the elite Marines—would dismay wartime commanders. Other mistakes, unknown and seemingly minor at the time, would cost American troops' limbs and lives. For example, the military's winter uniform was rated as protecting its user to a temperature of zero degrees Fahrenheit. This "spec" was fine for duty in Bremen or Japan but was deadly for service in the near-Arctic blast of the mountains of Korea's north. The army glove was not insulated enough to prevent frostbite—and was too thick to use while firing weapons.[50] Physically, logistically, and psychologically, America was unprepared for a "limited" land war in 1950. In the sort of global conflict the planners envisioned, Korea would be a footnote that the military believed would be unimportant, since U.S. airpower in Japan would be able to neutralize any Soviet threat from the peninsula.[51]

The view of Korea held by America's diplomatic infrastructure was even more complacent. In February of 1949 the CIA reported that the troop withdrawal would result in an invasion by the North and the fall of the South's government; the agency predicted, however, that the ROK army could eventually be built up to stand on its own. In response, the army issued its own report downgrading the probability of war on the peninsula.[52] The State Department's consensus of 1949 and early 1950 was that Korea was a political rather than a military problem. Diplomatically, the American position in Korea was tricky because the UN General Assembly had requested the withdrawal of *all* foreign troops, a measure with which the Russians complied. The State Department also worried that the presence of American troops would lead to the miring of America on the relatively unimportant peninsula. In sum, that there would be a *Korean* war, restricted to the peninsula itself, and limited in scope, was not in the post–World War II thinking of most military planners.

If such downgrades of Korea's importance had been kept secret, they might not have been registered as invitations to attack. But U.S. leaders broadcast signals that sounded as if they were uninterested in the fate of Korea. The first perceived green light pertained to the American definition of its "Pacific defense perimeter." The idea of a forward defense of America beyond its coastline was born early in the twentieth century in response to the development of modern transoceanic navies and the ascent of Japan. An early defense theorist and China adventurer, Homer Lea, warned that if Japan seized certain islands, such as the Philippines, then it would achieve a "possible supremacy of the Pacific littoral" that would give the empire "dominion over Asia and Asiatic hegemony."[53] In 1937, in a widely read book on geopolitical strategy, *The Struggle for the Pacific*, Gregory Bienstock argued for a defensive line that included Dutch Harbor in the Aleutians, Guam in the Mariana Islands, and Pearl Harbor.[54] After World War II, since America occupied Japan and had interests in China, MacArthur had argued that the "strategic boundaries of the United States were no longer along the western shores of North and South America; they lay along the eastern shores of the Asiatic continent."[55] Such a concept also lay at the heart of George F. Kennan's containment strategy against Communism, although he and MacArthur were to clash in the future.

By early 1950, however, there was no exact consensus as to which territories were the components of the perimeter, and Korea seemed to have been off the list. For example, in a January 12, 1950, speech to the National Press Club, Secretary of State Dean Acheson listed the Philippines and Japan within the perimeter, but made no mention of Formosa or Korea.[56] Likewise, Tom Connally, chairman of the Senate Armed Services Committee, stated in the spring of 1950 that South Korea was not defensible, and that the United States would end up abandoning it.[57] On June 19, days before the northern onslaught, John Foster Dulles lectured the South Korean National Assembly that the United States was "not bound by any treaty" to intervene to help the South.[58] Also, as noted in the visual analysis of the last chapter, *Time* magazine repeatedly failed to include Korea within the perimeter of American Pacific defense. These statements and actions were not intended to be direct invitations for a North Korean attack on the South, but they were interpreted by Communist leaders as confirmation of America's irresolution and unlikelihood to intervene if such an attack were to take place. Miscommunication was compounded, of course, by the lack of a diplomatic channel for "clarifications": there were no Americans in North Korea and there was no official American presence in the People's Republic of China.

When war came on June 25, America, although surprised, had a preexisting paradigm to fit the event into—an incorrect one, but one that seemed to jibe with the data available at the time. The North Korean attack, a manifest act of aggression, was assumed to be a Soviet plot, with Kim Il Sung (hardly known outside the peninsula) as errand boy, and China as lurking pawn. After all, the attack came only a few months after China had signed a treaty of alliance with the Soviet Union. Shortly before the onslaught, Secretary Marshall described China as being "literally under the direction of the Soviet Union."[59] This was the press consensus as well: as noted below in the analysis section, *Time*, until and even beyond the date of the Chinese intervention in Korea in November, projected in maps and diagrams, illustrated in photos, and described in captions and articles a worldview in which Russia was the controller of North Korea and that China was a conduit.

Other news sources concurred. A map in the July 3, 1950, isssue of *Newsweek* showed the Korean peninsula, Manchuria, and the tail of the Siberian Peninsula and Vladivostok. The topic of the map is the "Korean Cockpit" of North Korea's invasion of the South. Red arrows strike south from the north. The USSR is labeled, as is North Korea, South Korea, and even a section of Japan. There is no mention, however, that Manchuria belongs to the PRC. The caption reads, "Korean vs. Korean: How the Reds swept down." In *Time* of October 16, 1950, in the midst of an article titled "For a Free Korea," another map displays the same pattern. Titled "Korea's Waistland," it shows a more constricted view of North Korea and Manchuria. Here red-rimmed arrows show "Red supply lines" entering from Manchuria. Black arrows from the south mark the advance of the victorious American-UN armies toward "the Waist" (the narrower section of North Korea). Again, we see "Manchuria," but not "China" (see figure 2.1).

With such views dominating political and journalistic discourse, the public would be unlikely to hold an opposing view. Indeed, in December 1950, 81 percent of respondents to one poll said that the Soviet Union was responsible for the Korean War while only 5 percent thought the impetus for North Korea's attack came from China.[60] It followed that American assessments of Chinese behavior should be based on trying to gauge Soviet intentions.

At the same time, the Chinese had their own distorted view of American principles and capabilities. While the exact details of American geostrategic planning were unknown (and unfathomable) to the Communist leaders, the outward signs were of an America not willing to defend Korea. How war itself came out of this worldview is still a subject of controversy. But most scholars, and the weight of archival evidence, point to Kim Il Sung as the instigator of the war, not Stalin or Mao.[61] Apparently, it was Stalin who told Mao, during Mao's first Moscow visit, about Kim Il Sung's repeated requests for a green light to invade

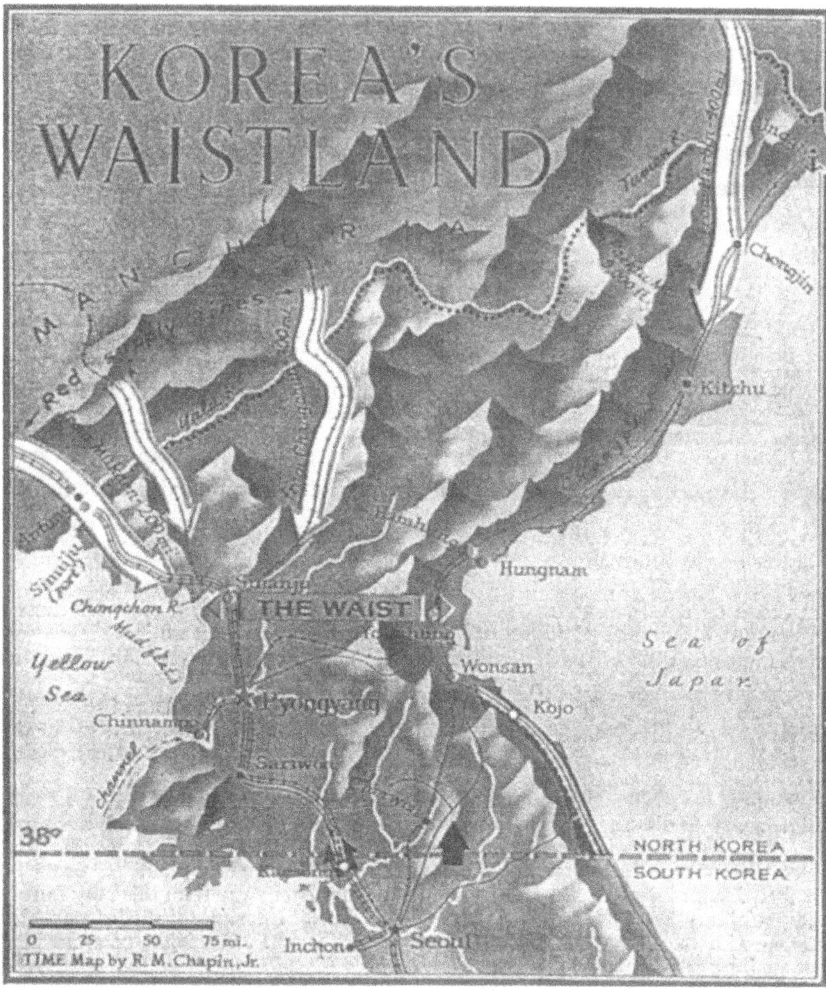

Figure 2.1. October 16, 1950: Map of "Korea's Waistland" (sic). (Artist: R. M Chapin Jr. Copyright 1950 by Time, Inc. Reprinted with permission.)

the South. Khrushchev, in his memoirs, describes Mao as being apprehensive at that time, unsure about the outcome of such boldness.[62] But according to Chinese marshal Nie Rongzhen (in his own memoirs), while Mao was still in Moscow a high-level North Korean delegation visited Beijing and asked the PRC to return fourteen thousand soldiers of Korean nationality who were in the People's Liberation Army. When notified, Mao approved the request.[63] (These arrived by June and were sent directly to the 38th parallel.) Both Chinese and Russian sources point to Kim as the driving element in the march to war.

But on September 15, the course of the war turned more decisively against the North. MacArthur, commander of all Allied forces in the theater, gambled and caught the North Koreans by surprise. American forces, mostly Marines, began landing at Inchon, near Seoul. The operation, code-named "Chromite," was one of history's great military maneuvers, "a masterpiece of amphibious ingenuity."[64] The harbor area was captured within days. Meanwhile, the Pusan force broke out and the North Korean position in the South collapsed. MacArthur was now faced with a choice: capture Seoul by the three-month anniversary of the initial North Korean attack—a great political coup—or drive east across the peninsula to cut off DPRK units fleeing Pusan. He chose Seoul. After hard fighting that reduced whole sections of the city to moonscapes, the general was able to announce on September 29, 1950, "By the grace of a merciful Providence our forces fighting under the standard of that greatest hope and inspiration of mankind, the United Nations, have liberated this ancient city of Seoul." Upwards of thirty thousand North Koreans were able to flee north, escaping entrapment.

3. Chinese Intervention

Now another difficult choice lay before the American administration: restore the status quo by pushing to the 38th parallel and halting there; allow ROK troops to drive north to attempt to finish off the DPRK forces and reunify the peninsula; or, most drastically, employ all U.S. and UN forces to "end the war" and crush North Korea. The military establishment, however, foresaw few impediments to a complete sweep of the peninsula and a substantial victory for anti-Communism that would overturn "the dangerous strategic trend in the Far East in the past twelve months."[65] Even veteran diplomats like Dulles argued that simply pushing back the North Koreans across the old border would create another threat that would have to be dealt with at some future time.[66] Yet he did offer a caveat, noting that America had "every reason to go beyond the 38th parallel except possibly one, and that is our incapacity to do so and the fact that the attempt might involve the U.S. much more deeply in a struggle on the Asiatic mainland with Soviet and Chinese Communist manpower because of the strategic bearing that the northern part of Korea has toward Port Arthur and Vladivostok."[67] MacArthur was the leading and most influential hawk, arguing as early as July that "it might be necessary to occupy all of Korea."[68] The Joint Chiefs of Staff were agreed that the United States should unite the peninsula, hold elections under UN auspices, and thus "displace part of the Soviet orbit." They did recommend, however, that a buffer zone should be created along the Chinese and Soviet frontier to "allay their suspicions."[69]

By September 1, even before the Inchon landings, Truman had decided to support the "right" of a unified Korea. Although the question would be raised again after the reconquest of Seoul, Truman issued a warning: "We hope in particular that the people of China will not be misled or forced into fighting against the United Nations and against the American people."[70] A key planning document, National Security Council 81.1, approved by Truman on September 11, urged unleashing MacArthur once he had established a beachhead, and also argued that neither Russia nor China was at all likely to intervene below the Yalu.[71] Thus, with most of his advisers in favor of continuing the war, Truman flew to Wake Island in the mid-Pacific to meet with MacArthur. The general was initially displeased at the request, complaining to the ambassador to Korea, John Muccio, that he had been "summoned for political reasons" and that Truman acted as if he was "not aware that I am still fighting a war."[72] The meeting itself went off better than he had anticipated. MacArthur assured Truman he had no political ambitions, and the two men got along on a personal level. Of the North Korean army MacArthur predicted that "formal resistance will end throughout North and South Korea by Thanksgiving."[73]

Of the possibility of Chinese intervention, MacArthur was dismissive to the point of being sympathetic with his possible enemy's plight. As described later by then special counsel to the president Charles Murphy, "the question came up about the possible intervention by the [Communist] Chinese and he [MacArthur] said they could not intervene effectively . . . if those poor Chinese tried to invade Korea, the part of Korea that we were undertaking to defend, it just made him sick to think of the way that they would be slaughtered and piled up."[74] In his own notes from the conference, Truman observed with his habitual Missouri plain-talk, "The Chinese Commies would not attack."[75]

Wake closed the deal. The president's postmeeting communiqué did not mention extending the war, but it did express the message that "We are fully aware of the dangers which lie ahead but we are confident we can surmount these dangers."[76] Most doubts swept away, the entire top administrative establishment approved MacArthur's drive north.[77] Nevertheless, the actual orders followed the guidelines of NSC 81.1 that advised caution against Chinese or Soviet moves: "make plans for the occupation of North Korea . . . if there was no indication or threat of the entry of Soviet or Chinese Communist elements in force."[78] The resulting official policy has been described as "run-and-stop": push north, but stay within South Korea's border, and halt if there is any intervention from China and Russia. Controversially, Truman included "secret instructions" to MacArthur on September 29 that held, "We want you to feel unhampered tactically and strategically to proceed north of the 38th parallel."[79]

The general's optimism is often cited as a failure of vision: he could not read Chinese signals, and dismissed their threats. Indeed, the traditional history of the

Korean War provides evidence for this scenario. As American, UN, and ROK troops plunged north, chasing the fleeing Korean People's Army, Communist China released a number of indications that danger lay ahead. But modern scholarship, especially the opening of some Chinese archives, suggests that these diplomatic signals must be reinterpreted. First, Inchon was a surprise to the PRC and the Soviets as well as the North Koreans, and recriminations and blame were pandemic. It was previously thought that the PRC warned the United States not to intervene and that the United States disregarded the admonitions; now we can claim that China told the United States that it *was* going to intervene to give Chinese forces time to deploy and the United States ignored the threat, perhaps purposively. The first warnings were statements by the Chinese leadership.

On June 27, 1950, Mao claimed that the U.S. administration had violated Truman's own pledges not to intervene in Taiwan, and in response, "Our people, and those of the whole world, should unite and be ready to smash any provocation by the American imperialists!"[80] The Taiwan linkage was an "open exposure by the United States of its imperialist face"; Mao further threatened that "United States aggression in Asia will only arouse the extensive and resolute resistance of the people of Asia."[81] Zhou Enlai echoed the chairman's sentiments, deeming the Seventh Fleet's deployment to Taiwan as "constitut[ing] armed aggression against territory of China and [is] a gross violation of the United Nations Charter."[82] Zhou also maintained, "That Taiwan is part of China will remain unchanged forever" and that "all the people of our country will certainly fight as one man and to the end to liberate Taiwan from the grasp of the U.S. aggressors."[83] Later, in a September 30 speech, Zhou Enlai stated, "The Chinese people absolutely will not tolerate foreign aggression, nor will they supinely tolerate seeing their neighbors being savagely invaded by imperialists."[84] That such rhetoric was ignored was not surprising. Chinese Communist hyperbole, by this time, was familiar to (and often dismissed by) the Truman administration. As shown in the analysis of *Time* magazine below, "Redspeak" was also regularly discounted by the American press.

A second type of menace came via diplomatic channels, especially the Indian ambassador to China, K. M. Pannikar. On October 3, Truman was told that Pannikar, and sources in his government, had passed on a message from Zhou that China would enter the war "if the U.N. forces, other than the ROK's, crossed the 38th parallel."[85] Unfortunately, practically no one in Washington trusted the Indian government of neutral-minded Prime Minister Nehru, which they viewed as naïve at best and supportive of the PRC at worst, or Pannikar, who was seen as "volatile and an unreliable reporter."[86] (As shown below, this assessment was visually and verbally reinforced in journalistic descriptions of Pannikar and Nehru at the time.) MacArthur for one did not understand why the Chinese would be so indirect with such important information.[87] Truman called

Pannikar's message "a bald attempt to blackmail the United Nations" and a transmission of "Communist propaganda."[88]

At the time, the American president could not see any benefits to China for an intervention and assumed that both China's and Korea's ultimate master, Stalin, had no interest in provoking a world war.[89] A CIA report of October 12 added weight to the dismissal of the Chinese threat by reporting, "The Chinese Communist ground forces, currently lacking requisite air and naval support, are capable of intervening effectively, but not necessarily decisively in the Korean conflict."[90] These views, when repeated, emphasized, and illustrated in the press, tended to find favor in the American public. A Gallup poll taken August 20 to 25 asked Americans to assess the chances of winning a war with Communist China. 20 percent said "excellent," 37 percent said "good," and 21 percent "fair," while only 11 percent measured the chances of victory as "poor." Surely viewing China as "lost" by American blunders or even traitorous foreign service officers rather than won by the CCP's political brilliance, and excellent tactics and troops, contributed to such complacency.

According to the Chinese viewpoint, of course, their megaphoned warnings appeared to have been ridiculed, ignored, or rejected. We also know now that Mao's threats were promises, not vague menaces. However reluctant he or others in the leadership might have been to engage in a war, the eventual conclusion was that China's security precluded an American-controlled country's bordering its richest province.[91] Mao was simply playing with America to gain time for his war preparations as much as in 1948–1949 he toyed with American officials to stave off a pro-Chiang intervention. Kim Il Sung had already asked China to intervene, and the evidence suggests that Stalin expressed support for this move and even offered to provide air cover for the PLA.[92] China had abandoned its Taiwan invasion plans after the initial DPRK attack, and Mao was converting the border army into a "Chinese People's Volunteer Army" whose mission was to "move into the Korean territory immediately to assist the Korean comrades in their fighting against the invaders until the realization of the glorious victory."[93]

Mao's anger was a factor here. His personality never embraced humility: the memoirs of his physician reveal a man who blamed all problems on others and who rarely accepted personal responsibility for any "mistakes" of state policy. He had been surprised by the quick and strong American response to the North Korean attack and Truman's linkage of Taiwan to Korea, having convinced himself and others in the leadership that America would offer only platitudes in support of the ROK. Mao's folly was, however, in his own mind and in his representations to his comrades, recast as confirmation of American perfidy. War was the only possible outcome. Mao now held that the American spearheads were aimed toward him and that a Chinese intervention was a defensive measure to head off an invasion of the motherland. As historian Chen Jian has argued, by the summer of 1950 Mao wanted war as part of his own domestic and international agenda, and little could have been done by anyone else to stop him.[94] Although Stalin incited and offered to assist him, the decision to intervene was basically Mao's alone for Mao's reasons alone.

The fact was, though, that by winter 1950, as MacArthur marched north, the Chinese poured south and, for whatever reasons, Stalin stood aside. In early November, the advance guard of the CPVA struck against an ROK corps and a few smaller American units in the East Unsan area of the North and killed or captured as many as twenty thousand South Korean soldiers. Nearby, the 8th Cavalry (a traditional moniker—this was actually a mechanized unit) Regiment of the 1st Cavalry Division was attacked and most of its positions were overrun by, as an official U.S. military history of the war described, "'human wave' assaults of bugle-blowing Chinese" so that within a day, "over eight hundred men of the 8th Cavalry were lost—almost one-third of the regiment's strength—in the initial attacks by massive Chinese forces, forces that only recently had been considered as existing only in rumor."[95] Yet within days, the Chinese disappeared. We know now that this was a tactical maneuver. The Chinese "volunteers" still awaited their full complement of armor and support troops and wanted to temporarily draw back closer to its supply depots. In addition, the Chinese military leadership followed its doctrine, developed from lessons learned fighting the American-armed Nationalists, of fight-and-retreat, which drew the enemy forces deeper in and masked the real Communist strength. The strategy worked: The allied forces crashed into well-entrenched and numerically superior Chinese armies, numbering over three hundred thousand men.

Within weeks panic replaced high hopes, and the military command foresaw a complete loss of Korea, with a retreat of the ROK to its own smaller Formosa and an offshore island named Cheju. On November 28, 1950, MacArthur reported to Truman and the Joint Chiefs that the allied forces faced "an entirely new war." The general asked for nuclear weapons, claiming that the war was lost without expanding the conflict to China, and began to criticize the administration even more loudly. In January, Seoul fell again to Communist forces. Truman, with the cooperation of the JCS, removed MacArthur on April 10, 1951, and his successor, General Eisenhower, was able to stabilize the situation. Korea became a minor seesaw of offensives and battles that decided the fate of a few dozen miles, north and south. Millions would die, but the war itself was a stalemate, and Truman would commit to no new gambles. Truce negotiations began on July 10, 1951, and would break down four times before an armistice agreement was signed on July 27, 1953.

Eisenhower and Peace

Most theories of decision-making in foreign affairs assign primary power to the office of the presidency—as indeed does the Constitution. Congress, special interest groups, foreign service bureaucracies, and even the press are seen in various theories as agents of influence over foreign policy in their own right. The "public," according to such analyses, is more led than leader, but often retains certain beliefs that politicians violate at their peril, for example, Don't get us into a war unless you really have to; don't spend a lot of money unless you prove it's worth spending. China policy was no exception. Americans, for almost the entire range of major involvement in Chinese affairs from the 1930s, consistently held certain basic positions: when asked, they paid attention to China; when pressed, they endorsed good relations with Nationalist China; after Korea they had negative feeling toward "Red" China; but they didn't support a war in China, neither in the 1930s to back China against Japan nor in the 1940s and 1950s to sustain Chiang against the Communists.

A Gallup poll taken between November 12 and 17, 1950, at about the time that the Chinese intervention in Korea was still in flux, asked if the respondents supported "the United Nations' forces [crossing] the border and fight[ing] them in China, or do you think the United Nations should keep the fighting in Korea only?" Almost half the sample supported "Korea only." On the other hand, 52 percent of the sample supported dropping atomic bombs on Chinese cities "if the United States does get into a war with Communist China" (45 percent were unambiguously affirmative; 7 percent said "yes, only as a last resort"). The median average of respondents, asked how long a war with China would take to win, said "four years." 57 percent of the sample, however, supported a deal whereby China would be given a seat on the UN Security Council if it would "agree to stop fighting in Korea." Another Gallup poll, taken between December 3 and 8, found 81 percent of the sample believing that "the Chinese entered the fighting in Korea on orders from Russia"; as shown below, this was the view the public was presented with by the press and the "experts" as well.

In 1951, Americans did not grow any more bellicose even after the full extent of China's intervention in Korea was known. A May Gallup special survey conducted of persons listed in *Who's Who in America* found 51 percent approved MacArthur's dismissal, while 46 percent disapproved. Fully 84 percent of the sample felt that America should not "start an all-out war with Communist China." A nationwide poll conducted between August 3 and 8 found that support for giving China a UN Security Council seat in exchange for peace in Korea had dropped; 60 percent opposed such a deal, while 27 percent approved. A poll conducted from February 4 to 9 asked Americans whether Russia was trying to manipulate the United States into an "all-out war with Communist

China." 79 percent agreed with this assertion. It is important to note that one of the most venerable of China visions—China's teeming millions—may have influenced the thinking of Americans. A September 29, 1950, poll had asked a national sample: "Just your best guess, what would you say the population of China is?" Only 12 percent answered correctly: between 400 and 499 million. 26 percent did not know, and another 26 percent put the number below 150 million. Americans may not have known exactly how many people lived in China, but they did believe that China had a huge army. When, in the February 1951 poll, they were asked to estimate, in the event of "all-out war," how many men China could raise for battle, almost half the sample listed "ten million or more" soldiers.

In short, Americans had cautionary feelings as well as antipathetic ones. The Truman administration was aware of this lack of groundswell for an "unlimited" war, and the strategic problem it represented had long been an object of government debate. George C. Marshall had commented in the 1940s that "America could not fight a seven-years war." A democracy, he estimated, cannot ask its citizens to fight an interminable foreign war not obviously in defense of its borders without prospect of victory and without visible progress toward closure. (Vietnam would prove him right.) Mao understood this; after the war and the confrontations with America over the Taiwan Straits in the 1950s he would comment, "Although imperialists will trigger wars, they are limited by pressure from the people."[96] In 1951, Truman comprehended that the situation in Korea was deadlocked; only supreme escalation, a huge commitment of manpower, the use of nuclear weapons, "unleashing Chiang," and bombing or even invading the Chinese mainland could create the opportunity for a traditional victory. The president, who once said that the proudest achievement of his life was that his artillery unit in World War I did not lose a single man in action, was simply not willing to gamble on a world war (as a worst case) or World War II-level American casualties (as a best case) in the name of a far-off peninsula in Asia.

In addition, the administration was well aware that China did not play by the same rules of popular will. Mao could dismiss the loss of millions, but Truman would bleed for the loss of thousands. On the one hand, as Evans Carlson observed in the early 1940s, the bond between officers and soldiers in the Communist armies was a strong one resembling parent-child affection and respect. At the same time, Mao's seeming unconcern about nuclear war was equaled by most Chinese Communist generals who did not figure battlefield deaths into their motivations, having been used to terrible losses in the Japanese war and the civil war.[97] For common American soldiers, this nonchalance translated to a seeming confirmation of the Chinese "mass" attacking in human waves, impervious to their own casualties. For American diplomacy, the problem was that Chinese leaders did not consider human costs of war an effective incentive to bargain in negotiations. Again Mao mused about the war: "Just how long will the fighting go on, and just when will the negotiations draw to a close? I say negotiations will continue, fighting will go on but there will be a truce. Why is it that there will be a truce? A thirty years' war or a hundred years' war is highly improbable, because a long war is very much against the interests of the United States."[98]

Meanwhile, by the spring of 1952, Truman was a beaten man—politically, if not in spirit—and had decided not to run for a third presidential term. His popularity, according to poll ratings, was as low as 26 percent, or roughly that of Richard Nixon when he resigned from office more than a decade later.[99] The Republican candidate for president seemed the perfect choice to replace him. As the former supreme commander of Allied forces in Europe, Dwight D. Eisenhower had immense prestige and the credentials to critique the Truman administration's stalemate in Korea. Candidate Eisenhower promised that he would "go to Korea" but offered no specific plan for either winning the war or ending it through negotiations. Furthermore, he argued that massive military spending that depressed America's civilian economy would hurt rather than help Cold War strategy. Obviously, a way "out of the Korean War" was a linchpin of such a policy. But Eisenhower seemed to understand that the American public's vision of China needed to be changed in a fundamental way. The American army, in both Chinese and American eyes, posed no current threat to China, or to the Soviet Union; that perception had to be reversed. (As discussed below, the American press cooperated closely in this propaganda effort.) The "stick" that prodded a conclusion to the peace talks needed to be a big one. It was also part of Eisenhower's campaign. One television ad featured "two soldiers . . . seen on a battlefield, discussing the uselessness of the war in Korea. Suddenly, one of the soldiers is killed; the other goes off, charging the enemy. The tag-line said simply, 'Vote Republican.'"[100]

The first step toward bolstering American standing and morale (at home and fighting the enemy in Korea) came in December, when, after his election landslide victory, Eisenhower traveled to the peninsula. Afterward he admitted to the press that there was "no panacea, no tricks" for ending the war.[101] But he exuded confidence that the situation was not doomed to stalemate and stasis. One incident highlighted his concern to dismiss the prevailing wisdom that the Chinese numerical superiority made combat victory impossible for the Allied forces. He joked to an enlisted man about whether he had seen any Chinese "hordes." The soldier replied in the same spirit, "I ran into a horde yesterday. I killed both of them."

Ike and John Foster Dulles, his secretary of state, held the basic philosophy that "Communists everywhere respect only force and hold fidelity to the pledged word in contempt."[102] While campaigning for president, Eisenhower argued that the United Nations could be a stronger force for world policing and that the United States should "build up a stronger system of all alliances with the nations of the free world, not only in Western Europe, but also . . . in Asia."[103] War in Asia, he reasoned, should be fought by "Asians against Asians." These positions played well with an American public wary of committing to an all-out war anywhere, but especially in Asia. The remarks were also threats to Mao, first to encircle China with enemies, and second to use Chiang Kai-shek's army either to invade the Chinese mainland or to intervene on the peninsula.[104]

In his State of the Union address on February 2, 1953, Eisenhower seemed to press home this latter menace when he announced that the Seventh Fleet would no longer "serve as a defensive arm of Communist China. . . . This order implies no aggressive intent on our part. But we certainly have no obligation to protect a nation fighting us in Korea."[105] This was a bluff, of course, because within days the Pentagon directed the Seventh Fleet to avoid any assistance to the Nationalist Chinese "in attacking the mainland."[106] Further, aid to Chiang was limited to material that could be used to defend the island, not enough to give the Nationalists any hope for a successful cross-channel invasion. (Luce's *Time*, as shown below, would advocate for more aid to support a real counterinvasion.) And China was listening. One major publication in the PRC denounced the Taiwan moves as a "harbinger of further Washington adventures in the Far East."[107] Later, Eisenhower concluded that the "unleashing" of Chiang, or at least the appearance of it, helped bring about the end of the Korean War.[108]

The last prong of Eisenhower's "peace" policy was the nuclear option. The president made a point of saying he would not restrain himself from using "the bomb" if suitable targets were found. This was not just a diplomatic maneuver. Eisenhower was serious, but on the other hand by that time copious United Nations command bombing had left few targets of any consequence standing in the North. Eisenhower did foresee dropping several nuclear weapons on major North Korean and Chinese troop formations, but he proved just as sensitive as Truman to the dangers of expanding the war and vetoed all suggestions to bomb targets on the Chinese border or in China.[109] A number of sources cite Mao as dismissing even the possibility of an American nuclear attack on Chinese positions in Korea or on the mainland. In any case, as discussed earlier, Mao's human calculus seemed not to reckon the loss of millions of his own citizens as tragic if it led to total defeat of the enemy. Mao's generals were less sanguine and less certain, but concluded that a new amphibious attack on the Korean coast was a more likely action by the father of the Normandy assault.[110] Soviet thinking was more sensitive, especially after Stalin's death and the inauguration of the Khrushchev era, in which outright war with the West was even less eagerly embraced as a policy option.

Until all the archives are opened we will not know how these measures of brinkmanship spurred closure of the war. In any case, a Korean armistice (but not a peace treaty) was signed by China and America on July 27, 1953, only months into Eisenhower's "talk tough" campaign. Neither side saw this agreement as the precursor of a long peace. Korea left America and China staring at each other across a great divide of ill feeling and aggrievement. As historian Chen Jian would sum up, "America's military intervention in Korea and China's rushing into a conflict with the United States finally buried any hope for a Sino-American accommodation, and the Cold War in Asia entered a new stage characterized by a total confrontation between the PRC and the United States that would last nearly twenty years."[111]

Evidence for such negative effects is found in basic survey numbers. A Gallup poll taken on June 2, 1950, before the Korean War started, found that fully 35 percent of the sample did not know that China had a Communist government. Of those who were aware of China's form of government, 40 percent opposed having ambassadorial relations with the PRC, while only 16 percent supported it. In contrast, in 1953, in a Gallup poll taken between May 30 and June 4, disapproval of giving the PRC a seat in the UN was solid at about 60 percent. Significantly, when the same question was asked again August 15–20, around the time of the finalization of truce talks in Korea, disapproval of a UN Security Council seat for China was up to 68 percent. Indeed, in the United States, negative views of China would dominate public opinion until Nixon's trip to the PRC in 1972 and, it could be argued, linger in some forms to the present day. As important, the China Lobby, even if it declined in potency (and actual membership) over time, found no real antagonist in the American public affairs debate: no "pro-PRC" constituency would exist in the United States until the explosion of trade in the 1980s. For his part, Secretary Dulles believed that new Communist aggressions would now be directed toward Southeast Asia.[112] This was an important mind set, possibly the basis of the initial involvement (and Lyndon Johnson's escalation) of the United States in Vietnam; this view held that all the struggles in Asia were parts of the greater one, and that all Communist aggressions were mapped out in Moscow and Beijing.

In China, one immediate by-product of the war was the campaign to fight bribery, tax evasion, fraud, and thievery, as well as economic espionage.[113] The main targets of attacks were the "patriotic capitalists," those business owners who had stayed in China after the revolution, believing Mao's slogans about a united front to rebuild China. With the campaign, many successful medium-sized and small businesses were destroyed or absorbed by the state. The property and funds confiscated gave the national budget a temporary boost, but the costs in long-term economic stability and growth were huge.[114] Mao, in assessing the expenses and benefits of the Korean War, understandably focused on what he perceived to be America's losses.[115] He claimed that the greatest benefit for China from the war was impeding an American invasion of the country, thereby preventing "a third world war."[116] This is still the official line today, although a tide of scholarship from within China and from Chinese scholars abroad is much more realistic in assessing the costs of the conflict. I, for one, have never met a Chinese—

manual laborer, student, political leader, or academic—who has expressed to me, privately, that the Korean War was of great benefit to *China.*

One major consequence for foreign affairs was little recognized at the time. Although Mao said to everyone, and probably believed, that the war was a great moral and political, if not geographic, victory, the conflict had sown seeds of doubt between the Soviet Union and China that would yield a malicious crop in the years to follow.[117] Stalin's reluctance to provide direct and immediate assistance to the Chinese in Korea, and his heavy-handed manipulation of his "allies," raised Mao's suspicions about the reliability of the Soviets.[118] Still, Stalin was Stalin the demigod; the birthplace of the World Revolution was still the USSR For Mao, outright rebellion was ideologically impossible. Indeed, from the civil war years onward, Chinese foreign policy was often more concerned with assuring loyalty to Stalin than objectively looking out for the interests of China. Mao would later claim that the Korean intervention was the decisive act that proved to Stalin that he was no Tito.

The death on March 5, 1953, of the "Great Helmsman" ended that allegiance. Ivan Arkhipov, the top Soviet economic adviser in China for most of the period from 1950 to 1958, believed that the chemistry between the two politburos changed with the passing of Stalin, and that by the end of the period, it was "above all the character of the leader of both countries" that destroyed the alliance.[119] Stalin's successor, Khrushchev, not impressed with Mao, the CCP, or confrontational doctrine, began to redefine Soviet security interests and military doctrines in light of nuclear weapons and their long-range delivery systems.[120] Increasingly, the two nations would find less reason to mask their differences, and more inclination to see an enemy in one another. By the end of the mid-1960s, China and Russia would come to the brink of war and, in several instances, actually fight battles along their border. The Korean War did not start or accelerate this decay of relations between Moscow and Beijing, but it was a crucial step in the longer process.

5. Pictures of an Unofficial War

Whatever ambiguities and uncertainties Henry Luce and most pro-China Americans of the right had in the year of China's revolution hardened into certainties during the Korean War. Anti-Communism returned as a triumphant and inflexible ideology in America, and for almost a decade to come the Lucepress would find nothing of value in the People's Republic of China (except the remnants of traditional culture). The Korean War also rejuvenated for Luce his loyalty to and hopes for Chiang Kai-shek and the Nationalist regime, which (a) scored its first victories on the long cross-channel struggles by defeating invasion attempts by the mainland, (b) began the slow process of undertaking much-needed reforms in its economic structures if not in its political ones, and (c) clearly was now America's most solid ally in Southeast Asia versus the threat of the alleged Sino-Soviet alliance. What marked Luce's feelings and Time*'s expressions about Taiwan was to emphasize the strong linkage between the need to support the Nationalist island as a stronghold in the defense perimeter against international Communist encroachment and attacks. Such feelings would lead Luce to some of his most extreme political advocacies, including agreeing with Douglas MacArthur's proposal to use Nationalist troops in Korea or in a direct invasion of the mainland, or even the use of nuclear weapons. If 1949 was the year of uncertainty in the Lucepress, the interlude of the Korean War and almost the entire 1950s marked absolute division between East and West, with little or no hope of reconciliation.*

The Korean War's body count belies any attempt to characterize the war as a comedy of errors, but we can look back and conclude that almost all parties involved blundered along, driven by false visions of their enemies, their allies, and themselves. Leaders of American warcraft and statecraft in 1950 visualized a China that was a toady to Stalin, too weak to be a real threat, a blowhard without real claws. Many images, generated by hopes and fears rather than cool analysis or actual inside knowledge, comprised such a mosaic. But there was also the factor of personal experience. The men who appraised China in the Pentagon and the State Department were not ignorant of China, but their views were limited by what they had actually experienced. For example, few living American military personnel had spent any time fighting or observing the Communists. The Pentagon based its estimation of the Chinese soldier and the Chinese general on the sorry relations with Chiang's often hapless legions and Keystone Cops-like commanders. The Communists' victories tended to be discounted as due to the deficiencies of their opponents, not their own discipline and innovative tactical ingenuity. Likewise, American generals such as MacArthur tended to fight in the same way that had previously brought them success in World War II.[121] This combination—relying on past strategies and treating the PLA of 1950 as if it were the Nationalist army of 1937—lay at the core of disappointment in Korea. As two historians of military error noted, "The failure of American leaders fully to understand that the enemy's situation and their own bore little resemblance to those they had faced less than a decade before best explains the debacle in North Korea."[122]

In looking at *Time* magazine's China pictures from 1950 to 1953 we gain a vivid sense of how such a false impression of China was illustrated for the home front. Whatever the reason, the visualization of China, or rather the picturing of the Chinas (the PRC and Taiwan), is an example of how our mental schema influence the pictures we draw of the world. The American media (and therefore the public) were unaware of the dangers involved since the political-military frame of mind was essentially replicated for public consumption. If Alexandre Mansourov is correct that, as noted above, "General MacArthur's repeated unconditional surrender demands, coupled with barely veiled direct threats against the PRC and the USSR, coming out of Tokyo headquarters, literally pushed the insecure Chinese to the brink," the "new" Chinese scholarship is likewise reliable in its assertion that Mao increasingly looked toward war as an expression of his policy and ideology. *Time* magazine, the most prominent American news publication, played an important role in taunting and threatening "Red" China, thereby provoking Mao to assume the worst about American intentions. If the American press could convince Mao in 1949 that the U.S. government was trying to sabotage the CCP from within, then the chairman perusing copies of *Time* from fall 1950 would be convinced that MacArthur's spearheads were aimed at him. Since such a vision of America confirmed Mao's malevolent picture of the United States, in an ironic sense one could say that the Lucepress and Mao were working hand in hand.

To begin the survey are a few notations about the overall China visions in *Time* of 1950 to 1953. The geographic foci by year, that is, the places where most pictures related to "China" were taken of or from were, respectively, PRC (1950), Korea (1951), PRC, Korea, Taiwan (1952), and Korea (1953). The most frequently pictured national identity of China-related persons was people from the Chinese mainland, mostly national leaders of the PRC or Chinese soldiers in Korea (after the intervention). In 1953, *Time* shifted focus to the Nationalists in Taiwan. For the entire period, the main topic of the images was war or military affairs. Another measurement taken in my coding of the images was the rate of social dysfunction, that is, negative events, from floods to civil unrest or famine, in the pictures of the PRC, or described by the captions and body text that referred to the pictures. The dysfunction rates, by year, were high, as usual: 1950 (65.5 percent), 1951 (77.8 percent), 1952 (54.5 percent), 1953 (68.7 percent). War, for each year, was the main social dysfunction, although there was an almost equal number of pictures of oppression and terror going on in "Red" China. In sum, the PRC was viewed very negatively by *Time* before, during, and after the period of the Korean War.

The magazine begins its coverage for the year 1950 fully focused on the Formosa debate. A foreign relations story examines a statement by President Truman indicating a reluctance to interfere in the Formosa issue. Twin political cartoons outline the polarities of U.S. policy options. One shows a plump Chinese military officer labeled "Chiang government" hanging at the very tip of a long branch labeled "Formosa." The subcaption reads, "The limb we are invited out on." In opposition is a picture of a porcine figure labeled "State Department" complete with spats and umbrella but with no head whatsoever, daintily holding a cup of tea, sitting atop a globe. He is holding a memo entitled, "We must condition the public mind to believe the fall of Formosa is not important." The caption reads, "Uh huh! Still saving face!" Yet the sequential subcaption of both pictures reveals *Time's* editorial view: "Appeasement is surrender . . . on the installment plan." At the moment when the Truman administration was trying to distance itself from Chiang, even possibly topple him, *Time* was attempting to rally its readers in his cause. In a further example, the issue also contains a close-up portrait of Madame Chiang Kai-shek, featured for a story about her trip to America to drum up support for Formosa. She is given more than half a column for a direct quote, calling upon America to help China because "ours is the cause of freedom." The following week, the endgame in China is reported as yet more "harassments and humiliations" of the United States. A central map shows a red curtain being dropped over China, Tibet, and other northern provinces. It is headlined "Lowering curtain," but the subcaption is "Imperialism." Obviously, what was meant was *Communists'* imperialism.

For almost a month thereafter, there were no China-related pictures in *Time*, but then in late February, there was a cartoon featuring Mao and Stalin looking at each other distrustfully. An accompanying article about the Sino-Russian treaty agreement and alliance (signed on the 14th) summed up, "It was the old, old story—the peace offensive to throw the Western world off its guard, cynical promises made to be broken, the indignant walk-out when the negotiations got down to specifics." Of the agreement, *Time* writes in reference to the Sinitic "mass": "The masters of seven hundred million people . . . seal[ed] their alliance with open panoply." Elsewhere in the issue, a profile of Nationalist general Sun Liren highlights improving morale and training of the army on Formosa. The implication is that the Nationalists are not done yet and should not be written off. Such a sentiment is driven home during the week of March 13 when the return to the presidency of Nationalist China by Chiang Kai-shek (after his resignation the previous year) is featured by two opposing images: the smiling Generalissimo and Missimo versus the dissident "acting" president Li Zongren and his wife in New York. The image of Li and his wife shows him with an even smile, but his wife imparts an expression of petulance. The story dismisses Li, saying it is unlikely that he will return to restore his authority in Formosa.

In the next few months, China was almost completely visually absent from *Time's* coverage, with the only real foreign news concerning some internal affairs in China and guerrilla activity in Vietnam. The Korean War first hit *Time* magazine via a map on July 3. The main focus of the stories is on the unpreparedness of the South Korean army; a sidebar notes, "[Korea] had virtually no military planes or tanks. Presumably somebody had thought that none would be needed." This is a reasonable assumption, but again the opening of archives and the making public of papers in the United States and elsewhere has suggested that the Korean army was not well supplied because the United States, although having had many warnings of an attack, did not *want* to give the fierce anti-Communist president of the South, Syngman Rhee, any means or incentive to provoke a war by attacking the North. *Time*'s issue includes a map of the western Pacific and East Asia. A second major feature of the map, one that will become a dominant cartographic motif until the Chinese intervention in November, is the ideologically driven labeling of what is "China." Here, China is shown bordering Korea, and Russia is not shown on the map. However, an unlabeled Manchuria is shown as an area *separate* from China. The story also assumes that the invasion was planned in Russia, and that China was either a coconspirator or simply a witness after the fact.

In the July 10 issue, "the battle of Korea" is featured in a map of the peninsula and of "Manchuria" in the north. The Russian role is visually emphasized in that the "supply lines" coming from Manchuria are labeled as being "Russian." In the following week, another map within a section on War in Asia, titled "Next?" displays most of the Eurasian continent, occluded by a Janus-faced Stalin from whom red-starred arrows point at possible targets, including Korea, Formosa, Indochina, Iran, Turkey, Yugoslavia, and West Germany. The story portrays the Korea invasion as a military gamble that was becoming a political failure for the Russians because of the unanimity and response in America and elsewhere in the West and the failure of the USSR to anticipate U.S. opposition and use its Security Council veto on any Korea related resolutions. (As noted above, recent evidence suggests that this absence was, in fact, purposive, to avoid setting up conditions to trigger the Sino-Soviet pact.) On July 24, *Time*'s War in Asia section featured a portrait of Chiang Kai-shek. The accompanying story is not wholly favorable, and indeed reads as a friend's chastisement. The Generalissimo is portrayed as being reclusive and stiff toward a continuing relationship with America but, on the other hand, having complete faith in MacArthur as the only "American who is capable of directing an effective U.S. program in Asia." The story advises the reader that the United States should work out satisfactory terms with Chiang and get him to rein in any nascent anti-Americanism.

A story on August 7 is a profile of Formosa which, while not fully laudatory, does hold that its army is in "top shape." Notably, the main weakness is deemed to be exactly what Chiang wanted to acquire in terms of equipment from the United States: modern fighter and bomber aircraft. This story again fits into a pattern of fundamental support for Chiang Kai-shek. On August 14, there is an all-is-forgiven image of General MacArthur during a visit to Formosa kissing the hand of Madame Chiang. *Time* takes some relish in noting that MacArthur's pro-Nationalist statements took by surprise and embarrassed the White House.

In the weeks to follow, there are only a few China images, most of them related to the issues of Hong Kong and guerrilla activity in Indochina. But then, on September 11, *Time* devotes an entire large section, including a dozen pictures, to a portrait of Formosa. It is almost wholly positive, as a simple recitation of the content suggests. First, there is a map of the whole island, which notably no longer includes an area designated as belonging to "headhunters." The map is linked to a report on "the land and the people" describing the name of Formosa ("beautiful isle" in Portuguese) and stories of the fierceness of the inhabitants in fighting foreign invaders. The "Portrait of a Land" pictures include the marketplace of Taipei, its grand hotel, the national defense headquarters, farmers and water buffaloes, billboards advertising U.S. films, a Japanese-built bridge across a river, pleasure boats, a teahouse on a cliff, children outside a farmhouse, a tenant farmer in his field, and, in the final picture, Nationalist soldiers training.

One image is a case study in the ideology of *Time*'s photojournalism. A wide-angle photograph shows a single, fit-looking young farmer in the foreground of a lush rice field on a balmy day. In the background, several other farmers till the land. The field is green and heavy with the harvest. The sky is azure blue. The caption informs us, "Rice fields like this one feed Taiwan's 8,500,000 people." In short, a few farmers and a few acres of land were used to stand for the entire island of Formosa. Based on the single example of visual evidence presented here, Formosa is (a) peaceful, (b) bountiful, and (c) self-sufficient in food. In all, Formosa is pictured as prospering under Nationalist rule, with even the Formosan natives "contented." This was factually not the case, but certainly conditions for the Formosans were better than they had been only a few years earlier. (The actual bitter divide between the émigré Chinese and the native Formosans will be explored in more detail in the next chapter.)

In the issue of October 9, a page-sized map of the Pacific and Eastern Asia is labeled "Asian Rimland." Formosa is shown almost at the apex of the U.S. circle of defense. The territory of Russia is raised to a cliff level and is signified not only by a giant "U.S.S.R." but also by a hammer and sickle which, graphically, is the epicenter of evil sending out red waves across Korea, Manchuria, China, Mongolia, and Tibet. As noted previously, *Time* had not included Korea in this defense perimeter in 1949. Also, at about this time Chinese units had begun to enter Korea in huge numbers and were readying to strike the U.S. and UN forces. Intelligence that this was happening was accurate, if not completely unambiguous. Yet *Time*, which had been even more dismissive of a Chinese threat to Korea, offers a picture of British-controlled Hong Kong and states that this may be the area of a new threat from Communist China. The 38th parallel argument is supplied in the story. Notably, in the listing of reasons to proceed north, several are offered to discount the Chinese menace: "Communist premier Zhou Enlai's threat that China 'will not stand aside should the imperialists wantonly invade' North Korea was only propaganda. If Russia or China intended to intervene in Korea, they should have done so earlier when they could have pushed U.N. forces into the sea." Another map, on October 16, shows Red supply lines coming from Manchuria without any reference to China (see figure 2.1), as does an October 23 map showing the "push for Pyongyang," and an October 30 map of North Korea showing the "flight" of the North Koreans, some into Manchuria. In every story, the prospect of a gigantic Chinese intervention is discounted and, indeed, China's proximity is hardly even mentioned. It is hard not to interpret this absence as a rhetorical decision by *Time*'s editors.

An internal image in the issue shows an Eastfoto photograph of a Beijing parade with hundreds of pictures of Chairman Mao being held up by the crowd. *Time*'s caption describes Mao as the "pudgy dictator." This is the kind of photograph-caption combination that would become familiar to American readers for almost two decades. Practically the only sources of pictures out of Communist China were Chinese news agencies such as Eastfoto. American editors were thus presented with a problem. Almost all of these pictures were, if judged by their content alone, positive propaganda for the PRC. The solution to the problem of countering the photos' original intent was such contrarian, mocking, or malicious captioning.

Time's hesitant and almost incredulous coverage of Chinese intervention in Korea is first indicated in a November 13 photograph of "Chinese Communist troops captured in Korea." A map shows the "battle of the border," where unlabeled arrows are pointing down, colliding with advancing UN columns. The story does mention that the Chinese have thrown back U.S. and UN forces, but this is not shown visually in any way, and the story does not admit that the Chinese are sending large numbers of soldiers into Korea, still speculating that there might be "ghost units" or that the Chinese are entering "piecemeal." Also of interest is *Time*'s description of U.S. and UN policy toward China during the war, as one of "turn the other cheek." This is obviously not the way that the Chinese viewed American intervention in their border region and in their border state of Korea. As late as November 20, another map of North Korea and "Manchuria" shows arrows bumping into each other but no pushing back of American or UN forces. Accompanying a story on the "Red mystery," the map asks whether the Chinese attacks are "bludgeon or blackmail?" with a gigantic question mark. The only combat picture related to China in the issue depicts American Marines, whose units at that time were surrounded by the Chinese, holding enemy prisoners.

The issue of November 27 continues to report confusion about Chinese actions in Korea. Was it a holding action? Why had the Chinese "failed to follow up their first stroke of intervention?" The story even ridicules worries that the Chinese would engage in a massive offensive to push back the UN forces from the peninsula. Again, this is at a moment when some three hundred thousand troops were in Korea waiting to meet the U.S., ROK, and UN spearheads. One may wonder whether *Time*, even if presented only with official information, might not have begun to ask questions about the military wisdom of the situation on behalf of its readers. A case in point: a derogatory cartoon shows a Siamese-twinned Mao Zedong split into a "Korea Mao" and a "Formosa Mao." On December 4, a map of the border region shows American troops still driving to the Yalu River, and a giant rope noose wrapped around parts of Manchuria and the remaining slice of land controlled by the North Korean army. The story asks, "Loose or noose?" It is not until the issue of December 11 that the full story of American and UN defeat is revealed. Here, a map is titled "Body Blow," with a graphic of a huge, spiked red fist smashing down from Manchuria into North Korea. The map also details the trapping of American units near the Chosin Reservoir and the drive of other forces backward to the South. The story describes "Chinese hordes" pouring down from the North. In the final issues of the year, other maps show the shrinking perimeter of the U.S. and UN defenses, with speculation about how the situation would be reversed.

Time begins 1951 with a retrospective on past U.S. relations with China. The history of the Korean War until then is also detailed. The story maintains that it was the Russians who instigated the North Koreans to attack, with no mention of the Chinese as being more than puppet accomplices. Included in the issue is a Sovfoto picture of a pro-Stalin parade in Beijing. The story offers an interpretation that the situation in Korea may represent a competition between Russia and China. The argument is that, while Russia was in charge of the initial North Korean invasion, China stepped in after the defeat of Kim Il Sung's forces and their headlong retreat north. Unfortunately, the sources listed for this speculation are "the U.N. mission in Korea" and Titoist officials in Yugoslavia, both of which would have been delighted if there had been a sign of disunity between Moscow and Beijing.

The weeks to follow in *Time* magazine are full of rhetorically unambiguous attacks on Truman's policy. A cartoon in the January 22 issue, for example, shows an Uncle Sam with one foot in a bucket of hot water representing Korea finding himself forced to dip his other foot in a bucket of hot water labeled "Defense of Europe." In battle news from Korea itself, an Eastfoto image of American soldiers near a jeep, arms raised in surrender to advancing Chinese troops, is featured. The caption reads, "Communist picture of surrendering U.S. troops." In the issue of February 5, in cartoons, the UN's inaction in support of Korea is mocked, but generally there is very little coverage of China-related aspects of the Korean War until February 19, when two images show, first, Korean Catholics praying for the "success of U.N. arms," and second, miserable-looking Communist troops who "kneel in mud of rice paddy [*sic*] and beg for mercy. They had been told they would be shot if captured." Other pictures from the early months of 1951 show Communist, that is, Eastfoto, images of captured Americans. One caption tells us, "Some GIs who escaped reported harsh treatment."

Time of April 23 suggests an "escape from stalemate." A map of western China, Korea, and Japan and parts of Indochina in the Philippines details MacArthur's plan for winning in Korea (the accompanying story calls MacArthur "the West's great success in Asia"). Among the elements of the map is most of the west coast of China from Hong Kong to Shanghai in flames in front of a Formosa labeled "potential fighting front." On May 21, *Time* shows a painting of Stalin and Mao sharing a moment of discussion in a book-filled study. *Time's* editorial view in relation to China is maintained in the issue of May 28, in which a picture of Chiang Kai-shek, his hands extended to make a point, is subcaptioned, "The authentic government of China." The accompanying story includes an argument for helping Chiang strike back at Communist China. Also added is an Eastfoto image showing China's leaders at the rostrum overlooking a May Day parade. The men are described as "Red China's bosses."

If there were any question about *Time's* position on identifying "China," it was put to rest by the cover of the issue of June 18. A not-unflattering paint portrait of Zhou Enlai with the outlines of pagoda-like buildings in the background is captioned, "Communist Zhou Enlai: The U.S.'s enemy is also China's enemy." Cartoons inside the magazine ridicule America's China policy but also criticize Republican congressmen who failed to decisively grill Dean Acheson during the hearing on China and Korean affairs. These twin iconographies are significant. *Time* magazine was not an instrument of monolithic propaganda. Tempting as it might have been for a critic of Luce to put it on a polarity of controlled press opposite Communist organs, there was undoubtedly nuance and complexity in the news reporting and visualization of China and Far Eastern affairs. To some extent, one sensed that *Time* respected power. Mao and Zhou are deemed "enemies" and "Red bosses," and the occasional epithet like "pudgy" sneaks through, but the men's successes are recorded with grudging awe, even if they are categorized as evil. Likewise, even on positions that Luce supported, the magazine was quite willing to attack its own allies if it felt they were poor spokesmen or had failed in their enterprises. This ambiguity is displayed illustratively in the June 25 issue in a map labeled "Red China." Graphic elements speak volumes about editorial intentions. Little men holding rifles designate the alleged position of "Nationalist guerrillas" in southern China; smokestack buildings designate "heavy industry"; and smiling skulls mark "terror centers" where, the extensive caption of the map describes, "the Communists stage their biggest mass executions of 'counterrevolutionaries.'" On the one hand, the reader is meant to infer that China is vulnerable to attack from Formosa from Chiang's forces. On the other hand, the reader cannot help but observe the iron grip of Communism throughout the whole region.

Another example of disdain for the PRC mixed with respect for the danger it poses appears on July 16, when *Time* shows a cartoon that displays a Chinese "gift horse" for peace, offered via Korea. In the story accompanying, there is an instructive quotation from Kipling's parable of the truce of the bear that is his warning about negotiating with the Russia of the previous regime, when the time the bear seems to be offering conciliation is when he is just preparing to strike. The negotiations that began at Kaesong at that time are in the weeks to follow portrayed as a "real struggle" and almost hopeless due to Communist intransigence. The rest of the year of 1951 has a very small number of China pictures, related mostly to trade and the truce negotiations (which invariably show the negotiators posed for a picture). But there is also a story on the corruption in Nationalist China and French battles in Indochina. One image appearing in the October 8 issue shows a "student rally in China" (courtesy of Eastfoto). The students each hold one fist in the air, and with the other hand hold propaganda posters and placards, while they face a speaker. The subcaption informs us that they are "ready to believe that black is white." The indoctrination of "promising teenagers" is described in the story as requiring "at least a year's brainwashing."

One of the last images of the year on China/Korea is in the issue of November 19 and is labeled "two-way threat." It shows the edge of a globe with China and the USSR at the top hovering menacingly over Japan; a giant arrow is thrust from the Soviet-controlled islands of northern Japan into the Japanese mainland. A final picture related to China is found on December 10, a portrait of a grieving Chinese American that illustrates the extortions of his people by the PRC. Again, as we shall see in the years to come, Luce's *Time* is focused on the plight (and on winning the hearts and minds) of the overseas Chinese community. Also, unambiguously positively *Time* never race-baits on the home front: American and overseas Chinese, with the exception of out-loud Communists, are almost always shown in a positive light.

The year 1952 is a light one for *Time*'s coverage of China-related issues, understandably so with the stalemate in Korea and the paucity of information coming out of China. One picture that did establish the view of what was going on in the PRC was the first of the year, in the January 7 issue. Titled "Red China Purge," an Eastfoto photograph shows trussed-up men with Chinese-language placards around their necks in a stadium being forced by soldiers to bow. The caption tells us that these are "counter-revolutionaries" who "received the death sentence from chanting crowds." *Time*, as usual, puts the information from the original Chinese caption into quotation marks, that is, "counter-revolutionaries," to cast doubt on its facticity. Another picture in the following week's issue is intriguing because today's viewer would see a striking symbolic connection to some of the most famous of pictures of China that would emerge thirty-seven years later at Tiananmen Square. A photograph shows Chinese men in uniforms exercising in what appears to be the field of a prison camp. The associated story describes how many Chinese prisoners in the war do not want to return to Communist China. The caption of the picture reads, "Red prisoners with homemade

Statue of Liberty. They feared the worst." Indeed, in the background is an approximation of the Statue of Liberty made seemingly out of papier mâché and wood; it holds up a torch and its face is vaguely Chinese. Powerful symbolism, the image fits in well with the ideology of *Time* and the American war effort: that the Red Chinese regime presided by terror over a people who were being brainwashed, but who, when exposed to the truth of freedom (via America), would rebel.

The ersatz Statue of Liberty, circa 1952, offers a contrarian footnote to this conclusion (see figure 2.2).

In the Tiananmen events of 1989, the Chinese government viewed the "Goddess of Democracy" that Chinese students erected then as evidence of foreign involvement in the protests. This was probably not true in the case of the plaster and Styrofoam figure constructed during the "Beijing Spring"; Chinese students designed it themselves and its meanings are more complex than those of a pure derivation of the Statue of Liberty in New York. But we do know that during the Korean War it was not a simplistic case of Communist POWs spontaneously yearning for the freedom of the West or Taiwan. They were provided

International

RED PRISONERS WITH HOMEMADE STATUE OF LIBERTY
They feared the worst.

Figure 2.2. January 21, 1952: Photo of "Ersatz Statue of Liberty Erected by Chinese POWs." (Artist unknown, "International Photo Agency.")

many comforts and a better diet than they had while in their armies, and they underwent education in anti-Communist principles; also, as Adam J. Zweiback documented,

> Converted Communist soldiers were often Nationalist Chinese sympathizers who had been conscripted by the Red army or Korean civilians swept up in the massive troop movements about the peninsula. Moreover . . . numerous [United Nations Command]-held soldiers were physically intimidated by both captors and fellow prisoners. Many leaders in anti-Communist compounds were actually veterans of Chiang Kai-shek's Kuomintang Army, and these agents administered frequent beatings to pro-Communist Chinese and North Koreans.[123]

More than fifty years later, we are unsure whether these "agents" got the idea, or were coached by American camp administrators, to erect a symbol that would cleverly serve the cause of nonrepatriation and anti-Communism.

Cartoons and pictures for the next several months focus on the Panmunjon peace talks, with the emphasis being on their glacial pace. A portrait of Chiang Kai-shek in the July 28 issue signals *Time's* full reestablishment of support for the Generalissimo. Although the article discusses some of the political and military mistakes he made in the past, and some of the problems in Formosa, the main argument is that although Formosa was improving in its situation, the island was not receiving enough American aid. The story relates, "Man for man, the Nationalist troops are in a fine state of training, well fed, well clothed and as pugnacious as terriers." The American reader would be forgiven for finding this to be a back handed compliment, albeit an unintentional one (can terriers win wars?). Indeed, this story, along with the smiling face of Chiang, reveals again one of the basic themes of this book: China visions are rarely monolithic. Just looking at an image like this, one would argue that the story was wholly positive. This is not the case. The story points out that Chiang's army is much smaller than the six-hundred-thousand-man figure often cited by proponents in the U.S. military who had wanted to employ it in attacking North Korea. (And that would be later cited by *Time* itself.) Overall, one gains the impression that Chiang can survive, but he would not be a decisive factor in a war with Communist China, nor could he hope to successfully reinvade the mainland.

The entire remaining visual coverage of China for the year 1952 includes only six images and marginal written coverage, with some speculation about Russo-China relations and the increasing guerrilla activity in Indochina. In short, for most of 1952, in *Time* magazine, China is out of sight, and those images that represent the PRC are almost completely negative or captioned to be so.

Time's wish to mold Formosa into an American image is shown in other ways as well. The first issue of 1953 features a photographic portrait of the America-based Chinese scholar, Dr. Hu Shih, and his visit to Formosa, which is described in the accompanying story as a "big, bright feather in the Nationalist cap." The story also establishes his anti-Communist credentials by noting that the "mainland Reds [have] spewed out a poisonous torrent of calumny against him." Shih compares the struggle of the Nationalists to that of the free French trying to shake off the German occupation. He also makes a mild call for more "freedom of debate and criticism in the press." One may wonder, especially when Luce's publications were characterized by his critics as being cheerleaders for Chiang, how such mixed messages arose from a *Time* magazine story. Luce's philosophy was not to simply laud what he liked and decry what he did not like in the world, but to try, in the style of a true missionary, to create a better world in the image to which his gospel called him. *Time* was not ignoring the problems of Formosa but was trying to nudge that country to be more America-like. Each criticism, however, was accompanied by a heavily underlined message, often illustrated, that over the straits lay an enemy with many greater sins and no virtues.

The February 9 issue of *Time* offers a smiling portrait of Chiang Kai-shek with the subcaption "Free to hit back." The article it illustrates focuses on Eisenhower's State of the Union address call to "unleash Chiang." From *Time's* editorial viewpoint, this is a liberation for Chiang and for *Time.* The story is optimistic in noting the strengths of Chiang's army and navy. Both this story and another that follows in the February 16 *Time,* which pictures Chiang Kai-shek in a car reviewing ranks of "Nationalist sailors," harp on the theme of "tools, not men," that is, the Nationalist army, which is now numbered at three hundred thousand, does not need soldiers but needs more equipment from the United States. In this story, we also see another notation that acts as a captioning of almost all pictures of China. A Chinese diplomat, the Nationalist delegate to the United Nations, speaks of the forthcoming liberation of "450 million Chinese." So does Dr. Shih in the earlier story from this year. It is interesting here, in these *Time* pieces, that the large population of China is used as evidence *in favor* of an attack upon it. The implication is that those 450 million are being oppressed by a small cohort of "Red bosses" and thus will be either inert or in support of a Nationalist assault on the mainland. Obviously, the argument could be put forth in the reverse: a nation of 450 million would be very difficult for an island of only a few million to conquer. In the same way, conversely, Mao would talk of China's millions as a reason why it did not fear nuclear war. Once again we see the mass of China being a sort of floating symbol for whatever outsiders and insiders of all political inclinations wanted to think of them, wanted to do to them, or wanted them to do.

Time's March 2 issue features an image of Taiwanese soldiers looking at a sand table map, presumably in preparation for military action, whether in defense (the map is of some part of Formosa or the Nationalist-held islands) or offense (the map is of a portion of the mainland) is not clear. The accompanying story announces that Chiang had "formally abrogated the 1945 treaty with Russia." This, the story explains, "clear[ed] the decks for action, [but] Formosa cannot undertake any major harassment of Red China—much less invasion—until she is much more heavily armed and equipped and better trained in amphibious operations and other techniques of war." The context not provided to the reader was that Chiang had had many dealings with Stalin in the past and, as explained in detail in the previous chapter, was not above "playing the barbarian against the barbarian," that is, working closely with the Soviet Union while trying to frighten the United States by asserting that the Red Chinese were Stalin's puppets. Another way one might put it is that for many years Chiang's anti-Communist ideology was clearly applied only to *Chinese* Communists. Now, pragmatically, he had correctly sensed that it would benefit his cause to be linked more closely to the international anti-Communist battle. As the *Time* story explains it, "After World War II, Russia backed the Communists in China, thus in effect making war on the nation which the treaty said was her friend and ally." Not only is this factually incorrect, it is hard to presume that, even with a limited knowledge of Sino-Soviet relations, the editors of *Time* believed it. As we have seen, until the last days of the Chinese civil war, Stalin retained good relations with Chiang, assisted the Chinese Communists minimally, often treated CCP leaders impolitely, and several times tried to hold them back from complete victory. In a single image and a few words, then, an important contextual history is falsified in the name of supporting Chiang in getting more "tools" to "clear the decks for action."

On March 23, *Time* features a photo of Madame Chiang Kai-shek sitting serenely among American politicians, including Vice President Richard Nixon, at a luncheon. The following week shows Madame Chiang returning to the airport in Formosa, a sure sign that *Time* deems all of her movements of significance. The pictures of the next few months deal with the peace negotiations in Panmunjon and with the continued fighting in Indochina. A pro-Nationalist army that has retreated into northern Burma is the subject of several feature articles. Then there is a blank spell of another couple of months, with only a few China-related images, associated mostly with Korea and Indochina.

In the October 12 issue, *Time* printed a picture of a crying Chinese prisoner "handed over to the Reds in Korea." Actually, the picture is from a scene in India, where, according to the truce terms of the Korean settlement, Chinese and North Korean prisoners were to be held in a "neutral" area for some time and then given the choice whether they wanted to repatriate to their homelands. The story casts doubt on Indian neutrality and describes the POWs as nearly all people who are pro-Nationalists. Several prisoners who are described as "possibly left behind as Communist plants" protested their treatment by the majority. Interestingly, the story explains that the "crying Chinese soldier" in the image was one of these pro-Communists. This fact makes the caption of the photograph problematic, because if the crying man is a Communist plant, indicating that he has been "handed over" implies unwillingness, as does the fact that he is crying. Again, *Time* has employed all the rhetorical devices of contrarian captioning, and indeed misrepresentation, to attack Communist China.

There are only a few more China-related pictures in *Time* in the lean year of 1953, but the final one, from December 7, sums up *Time*'s reportage and hope. On a reviewing stand, Chiang Kai-shek, smiling, in his characteristic cape and white gloves, stands next to a blank-faced Richard Nixon. The caption tells us that they are reviewing "combat maneuvers staged by 41,000 troops." The Korean War is over, but *Time* and, we assume, Henry Luce's hopes for China, have not stilled or gone silent. The focus on events in Indochina and the Indian border may be explained within this paradigm. China is no longer a hot war enemy/adversary, so the "liberation" of that country must look to some other flash point than a war in Korea. Chiang Kai-shek, the hero of *Time* narrative, now forgiven almost all his follies and failures, along with Eisenhower's envoy reviews the troops that will carry out the great enterprise. Perhaps the spark would come from Laos or Vietnam, perhaps elsewhere, but, as *Time* showed visually and in words in the year 1953, the dream of an America-ordered China lived on. In the years to follow, as we shall see, the confrontations between the United States and "Red" China would never lead to war, but would come close several times, and, in each case, America's shape-shifting of China would play a crucial role.

Notes

1. William Whitney Stueck, "American Policy toward China and Korea, 1947–1950" (Ph.D. diss., Brown University, 1977), 1.
2. Thomas J. Christensen, "Domestic Mobilization and International Conflict: Sino-American Relations in the 1950s" (Ph.D. diss., Columbia University, 1993), 300.
3. Tang Tsou, *America's Failure in China, 1941–50* (Chicago: University of Chicago Press, 1963), 529–31.
4. Nancy B. Tucker, *Patterns in the Dust: Chinese-American Relations and Recognition Controversy, 1949–1950* (New York: Columbia University Press, 1983), 190.
5. *Department of State Bulletin* XXII, no. 500 (January 16, 1950): 79.
6. Christensen, "Domestic Mobilization and International Conflict," 305.

7. Replies by the U.S. Department of State to a series of questions contained in House Resolution 452, 81st Cong., 2nd sess., Report of the House Committee on Foreign Affairs, February 9, 1950. United States Department of State, *American Foreign Policy, 1950–1955, Basic Documents*, vol. 2 (Washington, D.C.: Government Printing Office, 1957), 2456–58.

8. CQ, *China and the U.S. Far East Policy*, 1945–1967 (Washington, D.C.: Congressional Quarterly Service, 1967), 48. See also: J. H. Kalicki, *The Pattern of Sino-American Crises: Political-Military Interactions in the 1950s* (London: Cambridge University Press, 1975), 120–58.

9. Dean Acheson, *Present at the Creation* (New York: W. W. Norton, 1969), 306.

10. Bruce Cumings, *Korea's Place in the Sun: A Modern History* (New York: W. W. Norton, 1997), 258–59.

11. Robert Blum, *Drawing the Line: The Origin of American Containment Policy in East Asia* (New York: W. W. Norton, 1982), 182.

12. Cf. Zhai Qiang, "Recent Chinese Writings on 1945–1955 Sino-U.S. Relations," *SHAFR Newsletter* 20, no. 4 (1989): 75–84.

13. Zedong Mao, *The Current Situation and the Task of the Party in 1949. Mao Zedong Junshi Wenxuan-Neibuban (Selected Works of Mao Zedong on Military Affairs, Edition For Internal Circulation)* (Beijing: Jiefangjun Zhanshi Chubanshe, 1957), 1–3.

14. Michael W. Sheng, *Battling Western Imperialism: Mao, Stalin, and the United States* (Princeton, N.J.: Princeton University Press, 1997), 195.

15. Qiang Zhai, *China and the Vietnam Wars, 1950–1975* (Chapel Hill: University of North Carolina Press, 2000), 20.

16. NSC-68, "U.S. Objectives and Programs For National Security," in *Containment: Documents on American Policy and Strategy*, eds. Thomas H. Etzold and John Lewis Gaddis (New York: Columbia University Press, 1978), 409.

17. Byoung-Lo Philo Kim, *Two Koreas in Development* (New Brunswick, N.J.: Transaction, 1992), 169–70.

18. Bradley K. Martin, *Under the Loving Care of the Fatherly Leader: North Korea and the Kim Dynasty* (New York: St. Martin's Press, 2004), 87.

19. Joseph H. Alexander, Don Horan, and Norman C. Stahl, *The Battle History of the U.S. Marines: A Fellowship of Valor* (New York: Harper Perennial, 1999), 301.

20. Jonathan Pollack, "The Korean War and Sino-American Relations," in *Sino-American Relations, 1945–1955: A Joint Reassessment of A Critical Decade*, eds. Harry Harding and Ming Yuan (Wilmington, Del.: Scholarly Resources, 1989), 224.

21. Doris M. Condit, *History of the Office of the Secretary of Defense: The Test of War, 1953* (Washington, D.C.: Office of the Secretary of Defense), 2:228–342, quote found on 342.

22. Of extreme utility in describing the Korean War was: William Stueck, ed. the *Korean War in World History* (Lexington: University Press of Kentucky, 2004).

Also, unless otherwise cited, the following works were used to describe the build up and actions in the war: Roy C. Appleman, *Ridgway Duels for Korea* (College Station: Texas A&M University Press, 1990); Clay Blair, *The Forgotten War: America in Korea, 1950–53* (New York: Time Books, 1988); Gordon H. Chang, *Friends and Enemies: The United States, China and the Soviet Union, 1948–72* (Stanford, Calif.: Stanford University Press, 1990); Hao Yufan and Zhai Zhihai, "China's Decision to Enter the Korean War: History Revisited," *China Quarterly* 2, no. 121 (March 1990), 94–115; Chen Jian, "The Sino-Soviet Alliance and China's Entry into the Korean War" (working paper, Cold War History Project, Woodrow Wilson International Center for Scholars, Washington, D.C.: 1991).

23. Sergei N. Goncharov, John W. Lewis, and Xue Litai, *Uncertain Partners: Stalin, Mao, and the Korea War* (Stanford, Calif.: Stanford University Press, 1993), 215.

24. Dean Acheson, *The Korean War* (New York: W. W. Norton, 1971), 15–16.

25. Stueck, "American Policy toward China and Korea," 348.

26. Stueck, "American Policy toward China and Korea," 349.

27. Cumings, *Korea's Place in the Sun,* 255.

28. John L. Gaddis, *Strategies of Containment: A Critical Appraisal of Postwar American National Security Policy* (New York: Oxford University Press, 1982), chapter 4.

29. United States Department of State, *Foreign Relations of the United States: Diplomatic Papers, 1950,* vol. 7, *Korea* (Washington, D.C.: Government Printing Office),148–54.

30. Guoqiang Zheng, "Conflict, Contact, and Coexistence: Sino-American Relations, 1948–1960" (Ph.D. diss., University of Toledo, 1997), 113.

31. United States Department of State, *Diplomatic Papers,* 1950, vol. 7, *Korea* (Washington, D.C.: Government Printing Office), 348–50.

32. Zheng, "Conflict, Contact, and Coexistence," 121.

33. Stueck, "American Policy toward China and Korea," 356.

34. Stueck, "American Policy toward China and Korea," 358.

35. Anthony Leviero, "U.S. Not at War, President Asserts," *New York Times,* June 30, 1950, A1.

36. Stueck, "American Policy toward China and Korea," 357.

37. Stueck, "American Policy toward China and Korea," 357.

38. United States Department of State, *American Foreign Policy, 1950–1955, Basic Documents,* vol. 2 (Washington, D.C.: Government Printing Office, 1957), 2468.

39. Allen S. Whiting, *China Crosses the Yalu* (New York: Macmillan, 1960), 64–65.

40. I. F. Stone, *The Hidden History of the Korean War* (New York: Monthly Review Press, 1952).

41. Guoqiang Zheng, "Conflict, Contact, and Coexistence," 112.

42. Martin, *Under the Loving Care of the Fatherly Leader,* 62.

43. Martin, *Under the Loving Care of the Fatherly Leader,* 62.

44. Martin, *Under the Loving Care of the Fatherly Leader,* 62.

45. Roy E. Appleman, *1950: United States Army in the Korean War: South to the Naktong, North to the Yalu* (Washington, D.C.: U.S. Army, Center of Military History, 1992).

46. John J. Midgely, Jr., *Deadly Illusions: Army Policy for the Nuclear Battlefield* (Boulder, Colo.: Westview Press, 1986).

47. James M. Gavin, *War and Peace in the Space Age* (London: Hutchinson, 1959), 114.

48. David T. Fautua, "The 'Long Pull' Army: NSC-68, The Korean War, and the Creation of the Cold War U.S. Army," *Journal of Military History* 61, no. 1 (January 1997): 95

49. United States Department of State, *Foreign Relations of the United States: Diplomatic Papers, 1950,* vol. 1, *National Security Affairs, Foreign Economic Policy* (Washington, D.C.: Government Printing Office), 263, 282.

50. Jon T. Hoffman, *Chesty: The Story of Lt. General Lewis B. Puller* (New York: Random House, 2001), 384–85.

51. Alexander L. George and Richard Smoke, *Deterrence in American Foreign Policy: Theory and Practice* (New York: Columbia University Press, 1974), 73; T. R. Fehrenbach, *This Kind of War: A Study in Unpreparedness* (New York: Macmillan, 1963).

52. Jongsuk Chay, *Unequal Partners in Peace and War: The Republic of Korea and the United States, 1948–1953* (Westport, Conn.: Praeger, 2002), 124.

53. Homer Lea, *The Valor of Ignorance* (New York: Harper, 1909), 193–94.

54. Gregory Bienstock, *The Struggle For the Pacific: Port Washington* (New York: Kennikat Press, 1937), 53.

55. Thomas Etzold and John Lewis Gaddis, eds., "Conversations between General of the Army MacArthur and Mr. George F. Kennan," in *Containment: Documents on American Policy and Strategic, 1945–1950* (New York: Columbia University Press, 1978), 229.

56. Dean Acheson, *Present at the Creation,* 357.

57. Martin, *Under the Loving Care of the Fatherly Leader,* 67.

58. Martin, *Under the Loving Care of the Fatherly Leader,* 67.

59. Stueck, "American Policy toward China and Korea," 4.

60. Warren Cohen, "American Perceptions of China," in Michel Oksenberg and Robert B. Oxnam, eds., *Dragon and Eagle: U.S.-China Relations: Past and Future* (New York: Basic Books, 1978), 78.

61. Chinese Academy of Military Sciences, ed, *The PLA History of the War to Resist America and Aid Korea* (Beijing: Junshi Kexue Chubanshe, 1988).

62. Nikita S. Khrushchev, *Khrushchev Remembers,* trans. and eds. Edward Crankshaw and Strobe Talbott (Boston: Little, Brown, 1970), 368; Yufan Hao and Zhihai Zhai, "China's Decision to Enter the Korean War: History Revisited," *China Quarterly* 121 (1990): 100.

63. Guoqiang Zheng, "Conflict, Contact, and Coexistence," 120.

64. David Rees, *Korea: The Limited War* (New York: St. Martin's Press, 1964), 84.

65. United States Department of State, *Foreign Relations of the United States: Diplomatic Papers, 1950,* vol. 7, *Korea* (Washington, D.C.: Government Printing Office), 502–10.

66. United States Department of State, *Diplomatic Papers, 1950,* vol. 7, *Korea and China,* 386–87.

67. United States Department of State, *Diplomatic Papers, 1950,* vol. 7, *Korea and China,* 514.

68. United States Department of State, *Foreign Relations of the United States: Diplomatic Papers, 1950,* vol. 7, *Korea and China* (Washington, D.C.: Government Printing Office), 781.

69. United States Department of State, *Diplomatic Papers, 1950,* vol. 7, *Korea and China,* 502–10.

70. *Public Papers of the Presidents of the United States: Harry S. Truman, Containing the Public Messages, Speeches, and Statements of the President, 1945–1953,* (Truman address to

U. S. Congress, September 1, 1950 (Washington, D.C.: Government Printing Office), 609–14.

71. United States Department of State, *Diplomatic Papers, 1950*, vol. 7, *Korea and China*, 685–93.

72. Ambassador to Korea John Muccio, to John Wiltz, February 18, 1976. Miscellaneous Historical Documents Collection, www.trumanlibrary.org/wake/meeting.htm.

73. General of the Army Omar N. Bradley, chairman of the Joint Chiefs of Staff, substance of statements made at Wake Island Conference, October 15, 1950, compiled from notes kept by the conferees from Washington. Papers of George M. Elsey, 2, www.trumanlibrary.org/wake/wi198_2.htm.

74. Special Counsel to the President Charles Murphy, oral history interview, May 21, 1969, www.trumanlibrary.org/wake/meeting.htm.

75. Handwritten notes by the president, dated November 25, 1950, regarding the Wake Island Conference with General MacArthur. Papers of Harry S. Truman: President's Secretary's Files, www.trumanlibrary.org/wake/wi19_5.htm.

76. Statement by president Truman, October 15, 1950, pertaining to his meeting held on Wake Island with General MacArthur. Papers of George M. Elsey, www.trumanlibrary.org/wake/wi199_1.htm.

77. Omar N. Bradley, *A General's Life: An Autobiography* (New York: Simon & Schuster, 1983), 562.

78. Martin Lichterman, "Korea: Problems in Limited War," in *National Security in the Nuclear Age*, eds. Gordon B. Turner and Richard B. Challener (New York: Praeger, 1960), 31–56.

79. United States Department of State, *Diplomatic Papers, 1950*, vol. 7, *Korea and China* (Washington, D.C.: Government Printing Office), 792–93, 826.

80. *Zhonghua Renmin Gonghequo Duiwai Guanxi Wenjianji, 1949–1950* [Documents of foreign relations of the People's Republic of China, 1949–1950], trans. Shannon Wang. (Beijing: World Knowledge Publishing House, 1957), 130.

81. Yufan Hao, "Solving the Dilemma in China Policy: 1978–1979. A Case Study of Normalization of U.S.-China Relations and the Taiwan Relations Act" (Ph.D. diss., Johns Hopkins University, 1990), 33.

82. Hao, "Solving the Dilemma in China Policy," 33.

83. Hao, "Solving the Dilemma in China Policy," 33.

84. Chen Jian, *China's Road to the Korean War: The Making of the Sino-American Confrontation* (New York: Columbia University Press, 1990), 172.

85. Guoqiang Zheng, "Conflict, Contact, and Coexistence," 137.

86. United States Department of State, *Diplomatic Papers, 1950*, vol. 7, *Korea and China*, 869–73; M. Srinivas Chary, *The Eagle and the Peacock: U.S. Foreign Policy toward India Since Independence* (Westport, Conn.: Greenwood, 1995), 81–83.

87. Peter Lower, *Origins of the Korean War* (London: Longman, 1986), 192.

88. Harry S. Truman, *Memoirs* (New York: Doubleday, 1955/1956), 362.

89. Truman, *Memoirs*, 462–63.

90. United States Department of State, *Diplomatic Papers, 1950*, vol. 7, *Korea and China*, 933–34.

91. Xiaoming Zhang, "China and the Korean War: From a Chinese Perspective," *Journal of Asiatic Studies (Asea Yon'gu)* (South Korea) 41, no. 1 (1988): 63–80.

92. Guoqiang Zheng, "Conflict, Contact, and Coexistence," 145.

93. Jian Chen, "China's Road to the Korean War: A Critical Study of the Origins of Sino-American Confrontation, 1949–1950," (Ph.D. diss., Southern Illinois University, Carbondale, 1990), 210.

94. Jian Chen, *China's Road to the Korean War: The Making of the Sino-American Confrontation* (New York: Columbia University Press, 1990).

95. The Chinese Intervention, *U.S. Army. Korean War, Commemoration Brochures,* November 3, 1950–January 24, 1951, www.army.mil/Cmh-pg/brochures/kw-chinter/chinter.htm.

96. Tracy B. Strong and Helene Keyssar, *Right in Her Soul: The Life of Anne Louise Strong* (New York: Random House, 1983), 298–302.

97. Robert B. Rigg, *China's Fighting Hordes* (Harrisburg, Pa.: Military Service Publishing Company, 1951).

98. Zedong Mao, *Selected Works of Mao Tsetung,* vol. 5 (Beijing: Foreign Languages Press, n.d.), 79.

99. Cabell Phillips, *The Truman Presidency* (Baltimore: Penguin, 1969), 402–24.

100. Victor Kamber, *Poison Politics: Are Negative Campaigns Destroying Democracy?* (New York: Plenum, 1997), 30.

101. Dwight D. Eisenhower, *The White House Years,* vol. 1, *Mandate For Change, 1953–1956* (Garden City, N.Y.: Doubleday, 1963), 93–103.

102. Eisenhower, *The White House Years,* vol. 1, 123.

103. W. H. Lawrence, "Eisenhower Gives 10 Points for Peace," *New York Times,* September 5, 1962, 1.

104. *New York Times,* October 9, 1952.

105. Hou Tsung Chien, "U.S. Policy toward China, 1912 to Present Emphasizing Bureaucratic Level Analysis, Particularly Since 1936, of Both Nationalist and Communist Movements" (Ph.D. diss., Northern Arizona University, 1985), 218.

106. United States Department of State, *Foreign Relations of the United States: Diplomatic Papers, 1952–1954,* vol. 14, *China and Japan* (Washington, D.C.: Government Printing Office), 142–43.

107. Chien, "U.S. Policy toward China," 218.

108. Eisenhower, *The White House Years,* vol. 1, 123.

109. United States Department of State, *Foreign Relations of the United States: Diplomatic Papers, 1952–1954,* vol. 14, *Korea* (Washington, D.C.: Government Printing Office),170, 817–18, 836–37, 1067.

110. Yawei Liu, "The United States According to Mao Zedong: Chinese-American Relations, 1893–1976" (Ph.D. diss., Emory University, 1996), 272.

111. Jian Chen, *China's Road to the Korean War,* 1.

112. Stanley Karnow, *Vietnam: A History* (New York: Viking, 1983), 191.

113. Robert George Lee, "Revolution and Foreign Relations in the Thought of Mao Zedong" (Ph.D. diss., Brown University, 1980), 198.

114. Lee, "Revolution and Foreign Relations," 198.

115. Lee, "Revolution and Foreign Relations," 197.

116. Lee, "Revolution and Foreign Relations," 199.

117. Goncharov, Lewis, and Litai, *Uncertain Partners*, 222–23.

118. Goncharov, Lewis, and Litai, *Uncertain Partners*, 223.

119. Goncharov, Lewis, and Litai, *Uncertain Partners*, 222.

120. Goncharov, Lewis, and Litai, *Uncertain Partners*, 223.

121. R. F. Weigley, *The American Way of War: A History of United States Military Strategy and Policy* (New York: Macmillan, 1973), 387.

122. E. A. Cohen and J. Gooch, *Military Misfortunes: The Anatomy of Failure in War* (New York: Vintage, 1991), 192.

123. Adam J. Zweiback, "The 21 'Turncoat Gis': Nonrepatriations and the Political Culture of the Korean War," *The Historian 60*, no. 2 (1998): 345–62.

CHAPTER 3

Armageddon for Jinmen?: The Taiwan Straits Crises, 1954–1959

> The Chinese attack on the island of Jinmen set off "a sequence of events which was to extend through nine months, threaten a split between the U.S. and nearly all its allies, and seemingly carry the country to the edge of war, thus constituting one of the most serious problems of the first eighteen months of my administration."
>
> President Dwight David Eisenhower[1]

1. Introduction: China versus China, 1954–1959

By sheer volume alone, the average news reader of, say, 1950, was much more "informed" about China, visually and lexically, than his father or grandmother would have been. However, what the analysis of the pictorial coverage in *Time* of the last year of the Chinese civil war and the Korean War years has demonstrated is that attention, color, saturation, and focus via visual journalism do not generate organic understanding. Much of what *Time* showed about China was wrong, simplistic, or misleading. Such errors were not the product of purposive deception, but rather self-delusion and a journalistic ethic that demanded that the mantle of expertise and mask of sagacity about the events of the day be maintained at all costs. One yearns, in retrospect, for a *Time* editor—or for that matter Douglas MacArthur—to have said, "For God's sake, we know practically nothing about what's going in China, but let's try to look at the matter from their perspective." In short, in its eagerness to *shape* the news about China, *Time* failed to actually report the news about China.

The costs of this cracked mirror were not only the lives of American Marines at the Chosin Reservoir—the way America looked at China affected the domestic Chinese ethnic population. In the 1940s, when the government, Hollywood, and the press worked hard to valorize the Chinese people, by extension Chinese Americans benefited. Because moviegoers might weep for Wang Lung the farmer in *The Good Earth* did not mean, of course, that someone named "Wu" could get a job as an executive at General Motors, that a "Weilin" had any hope of being elected mayor of San Francisco, or that a "Sung" could become a member of a posh Los Angeles country club. But the conditions for Chinese Americans were certainly better than those of Japanese Americans. In the years during and after the Korean War, however, the situation reversed: China was then the blood enemy. The political climate of the early Cold War period placed Chinese Americans under intense scrutiny. As the Chinese American scholar Chiou-Ling Yeh has described,

> After 1949 Americans felt betrayed by the "fall" of China and made distinctions between Mao Zedong's "pagan and threatening" Communist China and Chiang Kai-shek's "Christian and supportive" capitalist China. China's involvement in the Korean War intensified resentment toward the Chinese. Mainstream media portrayed them as "inhuman," "treacherous," and "deceitful, fighting hordes" who were killing "our [American] boys," a drastic departure from the World War II–era image of Chinese as "heroic" allies.[2]

The response of the Chinese community, perhaps with the lesson of the Japanese internment experience as an incentive, was active and visual: The creation of extravagant rituals and spectacles, like the Chinese New Year's Day parade (first enacted February 15, 1953, in San Francisco), allowed Americans of Chinese descent to proclaim their loyalty to the United States and to popularize colorful and accessible parts of their cultural heritage.[3] Anti-Communism, often funded by donations from Nationalist coffers, was the trademark philosophy with which to exhibit American patriotism.[4]

Unlike the Japanese Americans, who were the victims of much scurrilous press coverage after Pearl Harbor, Chinese-American patriotism was lauded by powerful voices in journalism. *Time* magazine's visual analysis from 1954 to 1959 demonstrates and resonates an intriguing "pro-China" bias—the "other" China, that is. The unalloyed editorial line was that "Red" China was a terror state, powerful but oppressive. Although, as we have seen, the PRC controlled most of the photographic imagery about itself, *Time* undermined Communist propaganda with contrarian captioning. To *Time*, the overseas Chinese community—the diaspora from Hong Kong to New York—was a crucial "hearts and minds" battleground. *Time* often cited sources from this group, such as noted scholars, anti-Communist newspapers (based outside Formosa so as not to appear biased), and "ordinary Chinese," the proverbial fisherman or silk merchant, to justify anti-PRC stances. Stories also began to appear about the suffering of Chinese Americans who were blackmailed by the Communist authorities holding sway over their relatives in the motherland.

In addition, *Time* would publish stories set in the Communist present but would focus on the glories of dynastic arts and architecture. The implication was clear: what a shame that a regime of thugs ruled over a land of historical beauty and culture. The island of Formosa, the last bastion of the Nationalists, was, of course, the central symbol of the oppositional China. Interestingly, *Time* does not create a hagio geography. The country of Taiwan (as it has come to be known) is portrayed with some flaws: corruption, torpidity, scandals, and even anti-Americanism. There are even some references to aboriginal unrest—the native Formosans chafing at occupation by millions of mainland émigrés. But overall, Taiwan is the "good guy," the hope and beacon not just of the Chinese peoples but also of the world itself. In such an ideological climate, it seems less surprising that for the sake of a few islets—far from Formosa but held by the Nationalists—America twice almost went to war with China between 1954 and 1959.

2. The First Crisis

Formosa would seem an unlikely focal point for world crisis, with a population of eight million people in 1950 (twenty million today) and a surface area less than half the size of Scotland. It is hardly a lush paradise, a treasure of natural resources, or a strategically located region. Praising its remarkable economic achievements over the last fifty years, journalist Robert Kaplan called it a "barren rock on a typhoon-tossed sea." Until 1949, the island was a footnote in the history of East Asia. It was known to European traders and missionaries after its discovery by the Portuguese in 1590, who called it "Ihla Formosa" or "Beautiful Island." Formosa is itself an ancient land as well as a "melting pot" of peoples.[5] In 1624, when the Dutch attacked the island, they found little evidence of any Chinese control. They then set up a base on a small peninsula they called Tayouan ("Terrace Bay"), which would evolve into the word "Taiwan."

In 1637, the Ming emperor appointed a governor for Fujian Province and gave him control of all the offshore islands. These included Taiwan and the Pescadores but also the numerous rocks and islets near the coast such as the Tachens (to the north, about halfway between Formosa and Shanghai), Matzu (now known as Mazu), and Jinmen (meaning "Golden Door" or "Golden Gate" and also known as Kingmen, or Kinmen within the Amoy Bay no more than a few miles off the coast of the mainland. Yet the Dutch kept building up their franchise, and after defeating a Spanish force, they controlled the islands effectively from 1641 to 1661. Meanwhile, in China, the Manchus overthrew the Ming and established the Qing dynasty. In a foreshadowing of the Nationalist flight four hundred years later, Ming refugees and soldiers fled to Taiwan and set up a power base to fight against the Qing.[6] Some two decades later the Qing finally conquered Taiwan and restored it to the control of the governor of Fujian.[7] Although Taiwan remained an imperial backwater, the Qing, unlike their predecessors, allowed immigration from the mainland.[8]

The ethnic mix of the population, as of the dawn of the twentieth century, consisted of small aboriginal groups and a number of "Formosans," that is, people who had ethnic links to China but considered themselves, linguistically and culturally (and sometimes politically), to be an independent people. There was also a continuous migration of Chinese (Han) natives, mostly from Fujian Province. The local population was anything but happy with foreign rule: every few years they staged a rebellion. After the 1870s, in response to the Japanese takeover of the island of Okinawa, the Chinese took greater interest in their offshore possessions and improved and strengthened their administrative presence on the island. In 1886, Formosa became a province in its own right with its own governor. Then came disaster—the peace treaty of Shimoneseki, signed in April 1895 in settlement of the Sino-Japanese war, ceded the island to Japan. The entire native population rose in revolt when the terms were first made known. They variously invited other foreign states such as Britain and France to intervene on their behalf, only to meet refusal. All diplomatic options exhausted, the Taiwanese, on May 25, 1895, declared their independence, with the then Qing governor as the first president of the "Taiwan Democratic Republic." Notably, in self-description, they used the term "self-sustaining" (*tzu-li*) or "self-governing" (*tzu-chu*) rather than "independent" (*tu-li*)[9] and made clear that the decision was a maneuver to retain Chinese control over the island. But neither China nor any other state assisted in Taiwan's defense, and the Japanese army crushed local resistance by August 1895. The Taiwanese, noted for their spirit of protesting foreign rule, revolted again, and it was not until 1903 that local Japanese authorities would announce that the island had been fully pacified.[10]

The fifty years of Japanese occupation were enforced with an iron hand. Local culture, as well as Chinese influence, was suppressed; the inhabitants were forced to act as if they were quasi-Japanese citizens of the empire, which included the taking of a second, "Japanese," name and affirming reverence for the emperor. Those who offered opposition were imprisoned or assassinated. The island's agricultural economy developed as a "rice basket" for Japan and, starting in the 1930s, light industries were encouraged. The island was also exploited as a market for Japanese manufactured goods. At the same time, the Taiwanese were discriminated against within the legal system, barred from higher-end professions, and prohibited from demonstrations of native unity.[11] Still, the Taiwanese knew they fared better than other colonized peoples in Japan's empire, such as the Koreans or the Manchurians. The Japanese administration, while rigorous, was honest and efficient. Most fortunately, the island escaped the chaos and suffering of World War II and the Chinese civil war. As a result, Taiwan was economically and socially advanced beyond most regions of the mainland by the late 1930s.

Taiwan's diplomatic status followed the fortunes of war and politics. For Sun Yat-sen's and Chiang's Republic of China, the island was not an issue for quite some time. Several collections of the latter's speeches published in the late 1930s made no mention of it.[12] In 1937, however, when the second Sino-Japanese war officially began, the Taiwan issue arose again. On April 1, 1938, the GMD announced one of its war aims to be the liberation of Taiwan.[13] When China formally declared war on Japan on December 9, 1941, it abrogated all treaties between the two countries, including the Treaty of Shimoneseki. Immediately, in legal terms, as recognized by China and the United States and its allies, Taiwan was no longer a possession of the Japanese empire. This principle was articulated two years later at the Cairo Conference of November 1943, when the Western allies declared, "All the territories Japan has stolen from the Chinese, such as Manchuria, Formosa, and the Pescadores, should be returned to the Republic of China."[14] This commitment, with Russian approval, was reiterated at Potsdam in 1945, and on October 25, 1945, after Chinese troops had physically reclaimed the island, Nationalist China declared that Taiwan had been reincorporated into the country and that all of its residents were now citizens of China.[15] As noted in the previous chapter, as late as the spring of 1950 it was still the official position of the United States that, as Dean Acheson put it, "The Chinese have administered Formosa for four years. Neither the United States nor any other ally ever questioned that authority and that occupation; when Formosa was made a province of China nobody raised any lawyer's doubts about that."[16]

The post–World War II years were traumatic for the island's native population. At first, their feelings toward rule from Beijing were not wholly negative, although from the beginning there was a sense of separation and independence. Groups hoping for freedom from Japanese rule in Taiwan had been divided, some aspiring to reunification with China, some defining "freedom" as the right for self-determination for the island. After the start of the official war with Japan in 1941, the Nationalists would refer to the Taiwan question with the term "retrocession," that is, return to Chinese rule.[17] (The date the Nationalists retook the island would become the official holiday of Retrocession Day.) After reunification, however, there were political problems. The long years of Japanese occupation had made the Taiwanese suspect in the eyes of many mainlanders. A campaign to "de-Japanese-ify" Taiwan translated into outright prejudice against the native peoples.[18] Worse, the Nationalist administrators were singularly inept, callous, and venal individuals who managed within a few years to alienate large sectors of the local population. The most famous event that crystallized the new spirit of rebellion was the "2-28" incident (named because it took place on February 28, 1947), an uprising of islanders against the mainland authorities, especially the corrupt and harsh governor, Chen Yi. At first, the latter voiced conciliation, but only to play for time until more troops arrived, when he engaged in a wholesale massacre of the protesters. As many as twenty thousand were killed, and the uprising was quashed.[19] The incident remains to this day symbolic of the cleft between native Taiwanese and mainlanders.

Taiwan's situation drastically changed again in the last year of the civil war. Some two million mainland Nationalist Chinese fled to the island and Taipei was declared the "temporary" capital of the Republic of China.[20] (Chiang also left small garrisons on coastal islets such as Jinmen that technically were part of Fujian.) In the view of the native Taiwanese, the new mainlanders were in many ways an army and a constabulary of occupation. As we have seen, Taiwan's fortunes in this transitional period varied, to the point where the United States seriously considered toppling Chiang, but the Korean War solidified an alliance for two decades. Formosa was an island fortress in the chain of containment of Communism. On May 1, 1951, the United States set up Military Assistance Advisory Group (MAAG) in Taiwan to serve as an advising body for and as liaison to Nationalist forces. The United States also created several civilian organizations such as the Economic Stabilization Board for building a steady financial system that could sustain the island.

By the mid-1960s, U.S. economic and military aid to Taiwan had reached a total of $2.5 billion.[21] Taiwan was given preferential access to U.S. markets. Domestically, the Nationalists, after some stumbling, recognized the need for a new formula of rule, at least in economic policies. Farmers were given land when large estates were broken up. The rate of tenant farming decreased from 39 percent in 1950 to 15 percent by the late 1950s.[22] Industrialization and an export-oriented economy blossomed. Throughout the 1950s and '60s, the Taiwanese economic rate of growth regularly outpaced that of the mainland, ranging between 7.3 and 9.1 percent.[23] With this combination of relatively efficient and uncorrupt governance, political stability, and favored economic ties (and the exclusion of China as a competitor for the American market) Taiwan achieved its heralded economic "miracle."[24] At the same time, the GMD dominance lasted until the 1980s.

Chiang, although a refugee, was no wallflower. There were the repeated promises (or threats) by the Generalissimo to "liberate" China. In his New Year's message of 1954, Chiang announced that the coming year would see the liberation of the mainland. He was to rattle his rhetorical saber throughout the following months. From the early part of 1954 through the summer, a number of U.S. congressional leaders visited Formosa; all ended up making bellicose statements supporting the "liberation" of the mainland. China assumed that Nationalist attacks were orchestrated by the Military Assistance Advisory Group.[25] In 1953 and 1954, there were several clashes around the tiny Tachen Islands. The PRC captured several, but the United States convinced Chiang to defend the main islands of the group and backed up his defense with a presence from the Seventh Fleet. Finally, rumors of a mutual defense treaty of alliance between Washington and Taipei also fit into Mao's perception that an attack was being prepared against him, and that his own raising of tensions by bombing the Nationalist-controlled coastal islands would be a logical countermeasure. Indeed, as shown in the visual analysis below, *Time* magazine practically flaunted Chiang's "family album" closeness to the United States.

In sum, by the summer of 1954, Beijing's position was that "Taiwan had been converted into a U.S. colony as well as a military base for attacking our country; it was essential to put a stop once for all to the harassments of the traitorous Chiang Kai-shek group which, directed and assisted by the U.S. aggressors . . . [was] further reorganizing its . . . forces and blustering about preparations for an attack on the mainland."[26] In June and July 1954, there was another series of incidents that brought America and the PRC closer to war. First, China accused the United States of guiding Nationalist ships in their interception of cargo vessels, including Russian and Polish ships, headed for the mainland. Some PLA fighters shot down a Taiwanese passenger plane on July 23, and then during the rescue effort U.S. carrier-based fighters downed several PLA planes. In response, the PRC began a series of propaganda campaigns calling for the "liberation" of Taiwan. The rhetoric was just as strong from the other side of the straits. In visiting Taiwan in early August, Korean president Syngman Rhee called for the liberation of the mainland, with U.S. help.

International developments also conspired to precipitate a crisis. From April to July 1954 a major diplomatic conference took place in Geneva. The United States, the other Western allies, and the Soviet Union and China attended. The conference began days after the fall of Dien Bien Phu, a major French outpost in Indochina, to the Viet Minh forces. The outcome of the meeting was that independence was granted to Laos, Cambodia, and Vietnam, the last being partitioned "temporarily" between a northern Communist nation and a southern pro-Western country. Eisenhower, convinced by his hardline secretary of state John Foster Dulles, however, did not see any possibility of lasting peace brought about by any accord—the "domino" theory came into flower at this time, with the "head of the snake" (or dragon) being the PRC.[27] Indeed, Dulles advised the U.S. ambassador to France that the administration was considering active intervention in Indochina, whatever the result of the conference.[28] (By that time the

United States was underwriting 80 percent of the French war effort.) When an agreement was reached on July 21, 1954, the United States had walked away from the final negotiation sessions, and South Vietnam, on direction of the White House, refused to sign the final treaty. Dulles felt to participate at that point would be "prejudicing basic principles to which the United States must adhere if it is to be true to itself."[29] In a famous incident—which may have never actually occurred but that became the stuff of legend—Dulles even refused to shake Zhou Enlai's outstretched hand.

Practically, however, the offshore islets were the only targets of opportunity that the PLA could reach without challenging the American navy.[30] The "rocks" were not completely inconsequential in military terms because they provided an excellent lookout post for the Nationalists as well as staging areas for raids and intelligence operations and propaganda drops along the coast. Jinmen was important because no invasion of Taiwan was practical without first reducing or controlling the Bay of Amoy in which the small island lay. The harbor was one of China's best, and the United States had considered seizing Jinmen in contingency plans to retake Taiwan from Japan. Within a short time, then, at least four army-sized corps of the People's Liberation Army sat facing the straits. A considerable invasion transport fleet was assembled that, according to Nationalist sources, could have carried about a quarter of the PLA's troops at one time.

Now even more speculation began about what Mao wanted. A common explanation by anti-Communists in the West, which was offered repeatedly by *Time* magazine, was that China was engaging in border state provocations to distract its own people in times of internal conflict. Certainly no politician—Western or Eastern, Communist or capitalist—can ignore the fact that a foreign affairs crisis can be, at least in its early stage, useful for the purposes of focusing domestic political concerns and at least temporarily unifying disparate elements at home. But one major study of Chinese internal-external politics from 1949 to 1978 found very low correlation between chaos and disturbance at home and adventurism abroad, especially toward Taiwan.[31]

Social psychology must also play into any interpretation of China's behavior in the 1950s. First, Mao inherited a many-thousand-year tradition of the Middle Kingdom controlling directly or demanding vassalage from border countries and territories. In geopolitical terms, China wanted to be surrounded by minor powers that buffered it from foreign attack and that acknowledged the centrality and dominance of Chinese civilization. Second, Mao was a *Communist* revolutionary. He felt that the future belonged to him and his philosophy. As in the case with Kim Il Sung's proposal to allow the initial attack on South Korea, Mao felt that it was China's duty to assist anti-imperialist movements throughout the world, not least those of her neighbors. Mao was also a megalomaniacal egoist—his personal physician's memoirs leave no other character description possible. As in domestic policy, he could never leave any situation to be solved by others, but rather insisted on personal intervention and interference, always interposing his "great ideas." Many internal social, political, and economic catastrophes for China consequently resulted. Likewise, with evidence to the contrary, Mao believed himself to be a foreign affairs genius, able to gauge and predict the intentions and actions of other countries—like the United States. The crises of the 1950s can be interpreted with a view of Mao both practicing a pattern of "coercive diplomacy" and expressing an ego-driven need to assert primacy for China on its border areas.

Part of the context of this observation should be that Mao felt that his opponents constituted a brick wall. If Mao's personality was a driving force in Communist antagonism toward the United States, John Foster Dulles must be considered his counterpart in paranoia and intransigence. Dulles was a complicated man: he had opposed the dropping of the atomic bombs on Japan and felt that "moral authority" was the West's greatest weapon against Communism, but the Korean invasion had convinced him that a Manichean division in world affairs must be fought by all available weapons and that Communist treachery was infinite.[32] A study of speeches by a social scientist found that *any* message by the Soviet Union, antagonistic or conciliatory, was recast by Dulles as proving his image of Russia as an aggressive, belligerent archenemy.[33] (Dulles was, after all, the coiner of the term "brinkmanship" to describe his *own* strategy of international relations.)

Dying of cancer in 1959, Dulles's last public words about China were that recognition of the PRC would strike "a well-nigh mortal blow to the survival of the non-Communist governments of the Far East."[34] Dulles was no stand-back minister, either—a fact of consequence, since Ike's positions on the PRC tended to shift at times. In one Eisenhower biography, the president recalled that in times of crisis the two men would be talking, in person or on the phone, "eight or ten times" a day.[35] Reportedly, it was Dulles who personally convinced the president, on January 19, 1955, that Jinmen and Mazu should be defended, and then later, on March 6, Dulles further persuaded Eisenhower that an attack on the islands should provoke nuclear retaliation by the United States. In short, Mao's single-mindedness had in Dulles a backboard to hit with predictable response.

But Mao was not insane. During the civil war years he had developed a coherent philosophy for the need for a revolutionary army and its parent state to avoid pure passivity in war or in peace.[36] Among these "core doctrinal principles" was "active defense"—that is, "offensive defense, or defense through decisive engagements."[37] In the research of Alan S. Whiting, one of the fathers of the study of the Chinese military, these "principles of deterrence" fall into a pattern of behavior when a crisis looms in foreign affairs where military engagement is possible.

These include: "(1) clear demonstrations of a consistent signal, (2) coordination of military and political-diplomatic actions, (3) pauses to permit the opponent to respond favorably, and (4) efforts to prevent the opponent from escalating the conflict."[38] Whiting cites Mao's Korean War agenda as evidence of this pattern. He notes, for example, that the Chinese air force was under orders not to strike any targets anywhere near the 38th parallel. This school of what might be called Chinese "defensive aggression" has merit in interpreting Mao's actions in the later 1950s, but modern scholarship about the Korean War and the Chinese civil war tend to emphasize that Mao was much more of a gambler. Yet, in the first straits crisis (and even more so in the second), Mao was careful to avoid escalation. The Chinese air force was forbidden to directly attack any American ships, even when they were carrying Nationalist supplies or reinforcements.[39]

Meanwhile, as SEATO met, the PLA pounded Jinmen, the Nationalists were striking back, and hostilities were boiling on the coast of East Asia. America debated a response with the fundamental question: Was this worth a war? In Eisenhower's administration, hawks outnumbered doves. On September 12, the Joint Chiefs of Staff recommended that the president consider using nuclear weapons. Admiral Radford, chairman of the JCS, commented that, while the islets held little military value, surrendering them would deal "a political and psychological blow to the U.S. prestige in the Far East generally and would demoralize the Chinese Nationalists in Taiwan."[40] There was opposition within the U.S. administration, however. MacArthur's successor in Korea, General Matthew B. Ridgway, now army chief of staff, argued that fighting over islands so close to China's coast could not help but lead to direct conflict with the mainland.[41] Outright opposition came from Secretary of Defense Charles E. Wilson, who stated that the "doggoned little islands" could not be defended forever, either militarily or diplomatically.[42]

Eisenhower was of different minds about the situation, even with Dulles pushing him rightward. He did not want the American strategic focus to be diverted by the trivialities of the offshore islets, and he was seriously concerned that the United States would lose support from its allies if it tried to drive to war with China.[43] The president was also mindful that pubic opinion was less incendiary than some members of his staff. Between July 16 and 24, 1954, Gallup conducted a survey that asked a national sample of Americans: "Do you think the United States should be friendly to Red China and try to win her away from Russia—or should we treat Red China strictly as an enemy nation?" 47 percent answered "should be friendly"; 40 percent said "treat as enemy."

After the attacks on the offshores began, the public did not seem to rise in bloodlust. A September 16–21 Gallup poll, taken well into the firestorm over Jinmen, asked: "If Formosa is invaded by Communist China, which one of the following statements comes closest to your own view on what the United States should do?" The answers broke down as follows: Have U.S. planes bomb airfields and factories on the China mainland (10 percent); have U.S. planes and ships help to keep Communist China from invading Formosa (31 percent); have U.S. supply guns and other war materials but take no active part in the fighting (28 percent); have U.S. keep out of Formosa altogether and let them fight it out themselves (21 percent); no opinion (10 percent). In sum, Americans were not willing to go to war for Taipei, let alone for a rock in the Bay of Amoy.

Even without a popular mandate for war, the president wanted to sound tough. Eisenhower told the press, concerning nukes, "In any combat, where these things are used on strictly military targets and for strictly military purposes, I see no reason why they shouldn't be used."[44] Then, on December 2, 1954, the United States signed a mutual security pact with Taiwan, but U.S. assistance was promised only in response to an attack by the PRC on the main island or in the nearby Pescadores. The fifth article of the treaty asserted that any military strike "against the territories of either of the Parties would be dangerous to its own peace and safety and declares that it would act to meet the common danger in accordance with its constitutional processes." Article VII specified the right of the United States "to dispose [its] land, air and sea forces in and around Taiwan and the Pescadores as may be required for their defense." Within days, Zhou, speaking for the PRC leadership, attacked the new pact, which he called a "treaty of naked aggression."[45]

Move and countermove followed. In January 1955, the *People's Daily* called for other governments to join together to halt American "adventurist activities."[46] China then seized two small islands to the north, and the president asked from Congress and received the authorization for wider military latitude in the region.[47] The wording of the "Formosa resolution" gave the president the power to use force to protect Taiwan, the Pescadores, and "related positions and territories of that area now in friendly hands."[48] Eisenhower described the measure as a line in the seas that would "prevent the Communists from misjudging our firm purpose and national unity and readiness to fight."[49] Yet, within the "united" stand, there were footnotes hinting at the desires for de-escalation. The Tachens and a number of other small islands were evacuated. Further, the defense treaty with Taiwan removed the possibility of Chiang's making an independent invasion on the Chinese mainland that would provoke a war.

Another event that strengthened Communist China's international status was the conference of twenty-nine African and Asian nations, held at Bandung, Indonesia, April 19–24, 1955. Taiwan, along with South Africa, Israel, and the Koreas, was not invited. (The group expanded and eventually formed the so-called nonaligned movement in 1961.) The meeting was not intended to be a forum for attacking America. Indian prime minister Nehru, the initiator and organizer of the conference, selected the opening date because it coincided with the

180th anniversary of the Battle of Concord that had started the American Revolution. China, however, was now given a world stage free of American control or interference. Zhou spoke several times and assured the other countries that China had no aggressive intentions toward its neighbors.[50] At the invitation of the Indonesian prime minister, Zhou clarified his stance on the last day: "The Chinese people do not want to have a war with the United States of America. The Chinese Government is willing to sit down . . . to discuss the question of relaxing tension . . . in the Taiwan area."[51]

At Bandung, Zhou was able to articulate the "Five Principles of Peaceful Coexistence," whereby "different social systems" could be at peace and have normal relations. They included mutual respect for sovereignty and territorial integrity, mutual nonaggression, noninterference in the internal affairs of other states, equality for mutual benefit, and peaceful coexistence. Later, Zhou and Nehru would issue the principles as part of a joint communiqué. Their seeming mildness won China much respect—even though, in fact, an insistence on "mutual respect for sovereignty and territorial integrity" was, of course, a claim over Taiwan.

Having stood up to American power, interdicted any Nationalist invasion hopes, and won international repute as seeking "peaceful coexistence," China relaxed its pressure points. On May 1, 1955, the PLA stopped shelling all the offshore islands. The first crisis was over. The general restraint of both sides was displayed by there being only one actual instance of combat between American and Chinese forces: A U.S. plane was shot down, but the Seventh Fleet admitted it had been within China's three-mile limit and took no retaliatory action.

3. Talks and a Second Crisis

In 1955, Americans held no love for Communist China, an attitude that would carry on until Nixon's sojourn in 1972. In a survey conducted March 3–8, a Gallup sample was asked: "If we should get into a war with Red China, do you think we should use atom bombs?" 55 percent answered yes, 33 percent no. When asked the same question, but about "hydrogen bombs," support dropped to 44 percent and opposition rose to 43 percent. The Gallup pollsters noted that support for using atomic weaponry was greater among men than women, and among Americans over fifty rather than younger Americans. A special Gallup survey of May 25 asked a cross-section of *Who's Who in America* the question, "Would you approve or disapprove of admitting Communist China to the United Nations?" 34 percent approved and 61 percent disapproved. A national survey asking the same question May 12–17 found the disapproval rate at 67 percent. Another survey conducted from August 25 to 30 found the disapproval rate of China's admission to the UN at 71 percent. In a Gallup poll conducted

June 3–8, prompted by the first Taiwan Straits crisis, almost half the sample who answered the question "What do you regard as the biggest issue or problem facing the government today?" listed "foreign policy problems, working out a peace, dealing with Russia, Red China."

On the other side of the Pacific, amid the bellicosity of confrontation, Mao was moving toward some form of sustained communication with the United States. PRC diplomat Wang Bingnan recalled in his memoirs that Mao and Zhou felt that in 1954 China no longer faced any serious internal threats and could focus on improving international relations. The menace from America also seemed reduced. In Mao's worldview, the United States was overcommitted throughout the world, "invading" countries such as Japan, Korea, Vietnam, and Pakistan. The governments and citizens of these countries would soon revolt, he believed. There were pressing American reasons for some sort of talks as well. At that time, eleven U.S. Air Force fliers and forty-one American civilians were being held by the PRC. Eisenhower initiated the suggestion of secret talks to try to resolve the prisoner issue. To cover itself from an expected backlash of public opinion in the United States, especially from the China bloc, the administration described the meetings as being intended solely to help bring home the American captives.[52] After some give and take, the two sides agreed to ambassadorial-level talks, to be held in Geneva, with Wang as the PRC representative. The U.S. delegation was headed by Alexis Johnson, America's ambassador to Czechoslovakia. The day before the talks began, the PRC released the eleven airmen. Then, on September 10, 1955, the two nations signed the Agreement on Exchange of Civilians.

China perceived the agreement—on what for the Chinese was an issue of little consequence—as a step toward the negotiation of weightier matters.[53] Defense Minister Peng Dehuai (former commander of the "Volunteers" in Korea), opened the gate for expanded agreements when, on October 1 (the sixth anniversary of the founding of the PRC), he announced that China had no interest in continuing tension in the straits area.[54] A breakthrough remained possible, even when the United States tried to get Beijing to abandon the use of force to settle the Taiwan issue. By May 1956, a Chinese draft of a "nonuse of force" agreement seemed acceptable to both sides:

> The People's Republic of China and the United States of America are determined that they should settle disputes between their two countries in the Taiwan area through peaceful negotiations without resorting to the threat or use of force against each other.[55]

The Chinese precondition, however, was that any final agreement should be negotiated in person by Dulles and Zhou. The former could not bring himself to trust the Communist government and had no interest in a face-to-face meeting with his counterpart, thinking such a meeting would only undermine the American position in Asia.[56] Moreover, this was not 1972, and as public opinion scholar Melvin Small points out, "No one could have been elected to a position of power in 1956 who talked openly about sitting down with Mao Zedong"—the latter whose hands, in the eyes of Americans, were dripping with the blood of innocents.[57]

But for Zhou, a high-level meeting was a crucial indicator that America was serious and would accord China "respect" as articulated in the "five principles."[58] In response to the stone wall, which must have reminded the CCP leaders of their old nemesis Patrick Hurley, Beijing tried to go around Dulles in several ways. First, Wang was instructed to try to cultivate a friendship with U.S. Representative Johnson.[59] Second, the PRC started granting visas to U.S. news personnel and inviting foreign news agencies to send journalists to cover what was going on in China (albeit under heavy censorship).[60] It was the United States that kept the ban in force, citing that there were still Americans being held as "political hostages" in China. To the Chinese such a reason seemed contrived.

At the same time, Beijing tried to conciliate Taiwan itself. In September 1956, Liu Shaoqi, later to become president of the PRC, stated at a Beijing conference that "the liberation of Taiwan is a complete internal problem of China. We are willing to use peace talks rather than armed force to bring Taiwan back into the arms of motherland. We will have to use force only when all possibilities of peaceful negotiation are all lost or the peace talks all fail."[61] Mao even held off on deployment of new air units to Fujian and let it be known via backdoor channels to Chiang that the PRC was willing to come to some accord whereby he still retained his armed forces while the province was legally returned to China.[62] But the words and actions from Washington remained tough. On March 12, 1957, Dulles argued that UN membership for Communist China would simply "encourage influences hostile to us and to our allies."[63] The talks finally died under controversial circumstances. In December 1957, after some seventy-three meetings, the United States sent a lower-level representative instead of Ambassador Johnson, but China insisted on continuing talks only at the ambassadorial level.

The failure of negotiations did not immediately lead to another round of fighting, but certainly the clock was ticking. Once again, as well, Chiang was sounding his own war gong, much to Beijing's irritation. ROC troop strength on Jinmen and Mazu increased to about one hundred thousand. The Nationalist air force patrolled the mainland and attacked military installations on the coast. Sabotage, commando operations, propaganda drops, and attempts to foster local resistance groups also continued.[64] Mao suspected that Chiang Kai-shek wished to launch an invasion of the mainland before Eisenhower left office, and felt that renewed PRC pressure on the coastal islands would warn Washington against supporting any Nationalist adventures.[65] A leading Chinese newspaper proclaimed that the United States was "speeding up its preparations for a nuclear war against China."[66] There was also a rumor that the United States planned to deploy tactical nuclear missiles in Taiwan by May 1957. In fact, Ike did deploy nuclear weapons to Taiwan in January 1958.[67] These were Matador (MGM-1) nuclear missiles with a 690-mile range (the farthest-reaching missile in the U.S. arsenal). At the time, the operation was secret and was revealed only through declassified documents years later.

The PRC made its own moves. In August 1958, Nationalist sources reported that hundreds of Soviet-built Chinese MiGs had redeployed to air bases along the coast, with the greatest concentration near the straits.[68] In addition, the PLA strengthened its artillery batteries in Amoy Bay, facing Jinmen Island. The number of local PLA land units increased as well. Other factors were at work. Mao saw an excuse for action via "social international solidarity." In the American intervention in Lebanon in July 1958, the CCP leadership linked "imperialistic" aggression in the Levant and Taiwan and set off another anti-America and "free Taiwan" propaganda campaign. On July 17, 1958, the mayor of Beijing announced to the crowd at an anti-imperialist rally, "At present the U.S. imperialists still occupy by force our territory of Taiwan. We Chinese people are determined to liberate Taiwan and have full confidence that we will achieve this."[69] But it soon became apparent to the leadership that it was hard to sustain the prospect of a war in East Asia rooted in a crisis in the Middle East. Mao put off an attack so as to deal with "Chiang directly and the Americans indirectly."[70]

The second straits fighting officially began on August 23, 1958, with a massive shelling of Jinmen. As many as forty-one thousand rounds hit the islet within the first few hours. The onslaught coincided with a welcoming ceremony for a tour by the Nationalist defense minister. He escaped unharmed, but a number of senior commanders were killed. Fire was concentrated on Jinmen's airstrip and landing zones, and the next day Beijing Radio announced that a landing assault on the islands was imminent. At the same time, the Chinese made an appeal to possible war-adverse elements in Taiwan to back down from a death struggle for the small coastal islands. Defense Minister Peng, in a radio broadcast aimed at the Taiwanese leaders and population, appealed: "The Americans will surely abandon you some day. Don't you think? The muse of history will prove this to you . . . . In your position, can you feel the blood running cold? The American imperialist is our common enemy."[71] A few weeks later, Peng broadcast,

> Friends of Kuomintang: do you not sense this danger? Where is the way out?—Of course, we are not asking you to break with the Americans right away. That would be an unrealistic idea. We only hope

> that you will not yield to American pressure, submit to their every whim and will lose your sovereign rights, and so finally be deprived of shelter in the world and thrown into the sea. These words of ours are well-intentioned and bear no ill-will. You will come to understand them by and by.[72]

And the rain of fire continued. Over the month of August alone the PLA slammed 571,959 rounds of shells on the island.[73] (A veteran of the defenders related that among his comrades the running joke was that for years Jinmen would be the richest metal mine in the world.) The ROC F86s, armed with Sidewinder missiles, shot down 29 MiGs in the first month of the fighting.[74] Soon, however, the garrison ran low on ammunition, and the ROC navy could not reliably or safely penetrate PRC shell fire and air cover.

For many in the American leadership the second crisis had simple roots: a Moscow-Beijing plot. Eisenhower, in his autobiography, although he calls speculation about Sino-Soviet differences "intriguing," generally concluded that the Russians must have had a hand in provoking both straits crises, especially that of 1958, because "he [Khrushchev] may have been rankled because we had ignored his threats on Lebanon, and . . . concluded that a reopening of the offshore islands issue might divert the attention of the world from Lebanon to the Far East . . . [and] to test Western unity and resolve—a place where, in the light of divergent viewpoints among Western allies, the Communists might exploit apprehensions regarding Soviet advances in weaponry."[75] The main piece of evidence for this conclusion was that a visit to Beijing by Khrushchev preceded by a short time the opening of hostilities by China. Indirectly, the United States also perceived that the Jinmen attack was well timed with Soviet advances in the arms race such as the launching of *Sputnik.* Eisenhower commented in his autobiography, "The new situation included new dangers that seem to make our position more difficult."[76] It made sense that, emboldened by its new powers, the Communist bloc was testing the West yet again.

On the other hand, there is strong evidence Mao incited the second crisis without Moscow's knowledge. The Soviet Union did make strong statements in support of China once the crisis had begun, but it was noted that these did not actually occur until the situation had already begun to defuse, and there was little threat that Moscow would have to take any concrete action in support of the PRC.[77] The timing of Khrushchev's trip, as well, may have been coincidental. Reports of the talks between Khrushchev and Chinese officials made no mention of Taiwan and speculation still exists about whether the issue even arose.[78] Mao himself reported later that the discussion "did not contain a word about the question of the Taiwan situation."[79] A Soviet commentator also confirmed that Mao may have made the decision to attack Jinmen again in 1958, but he did not mention it to Khrushchev during their meetings.[80] There is also evidence that the second straits crisis may have been initiated by the PRC in part to torpedo Khrushchev's "opening up" to the West. Diplomat Wang Bingnan reports in his memoirs that Mao wanted to embarrass Khrushchev, who he felt was conceding too much to the United States. The memoirs of Li Zhisui, Mao's private physician, also support a Mao who wanted to upstage the Soviet premier, not consult and obey him.[81]

In the first days of September, the United States decided to help break the blockade, and President Eisenhower and Secretary of State Dulles made repeated guarantees for the safety and security of Taiwan. These assurances were backed up by the delivery of a number of advanced weapons systems, including ground-to-air missiles, and the conducting of joint war games on Taiwan with Nationalist forces. September 4 proved to be a crucial day. The People's Republic of China announced that the three-mile limit of sovereignty of the sea would now be extended to a twelve-mile territorial limit which would, naturally, contain all the small offshore islands under contest. Dulles met the same day with Eisenhower at Newport, Rhode Island. At the subsequent press conference, Dulles talked tough about how the "naked use of force" against the offshore islets would "forecast a widespread use of force in the Far East which would endanger vital free world positions and the security of the United States." But he closed with a call for a peaceful resolution to the crisis and made a specific reference to the Geneva discussions as a possible vehicle for resolution of the crisis.[82]

Within days, Zhou Enlai also spoke in favor of a renewal of ambassadorial talks, and Eisenhower quickly agreed. U.S. ships escorted supply convoys starting on September 7, 1958, but only to the three-mile limit near the islets. Finally, the ambassadorial talks restarted on September 15, this time in Warsaw.[83] In October, Dulles visited Taiwan; his joint communiqué with Chiang stated that the ROC would not try to invade the mainland; the Jinmen garrison would be drawn down in size; but the United States would provide Taiwan with even more sophisticated weaponry.[84] The Warsaw talks would continue sporadically, totaling some 136 meetings before they came to an end in February 1970.

Mao showed his intentions to go the direction of negotiations in other ways. He held the PLA to "cut off without killing" the supply effort from Taiwan.[85] The PLA air force was restricted to flying within the newly claimed twelve-mile territorial sovereignty.[86] On October 6, Defense Minister Peng Dehuai announced a halt of the shelling for a week, knowing full well that the respite would give the United States time to shuttle materials and reinforcements to the beleaguered islet. On October 13 he suspended shelling for two weeks, then a few weeks later he initiated the infamous "odd/even" tactic, shelling the islet only on even-numbered days. Mao commented that "militarily it sounded like a joke, since such policy was unknown in the history of Chinese or world warfare. However, we are engaged in a political battle."[87] Eisenhower's reaction was to call the developments evidence of "a Gilbert and Sullivan" war.[88] Mao also gave a series of orders to minimize the possibility of any great number of Americans being killed, and to avoid shooting at American ships directly.

On the American side, there were also strong political reasons for restraint. The United States' diplomatic position during the 1958 crisis had eroded considerably from the previous confrontation. In John Foster Dulles's estimation, the United States Security Council veto was practically the only block to Chinese admission to the UN.[89] The situation had changed domestically in 1958, as well. Democrats began to assert divergence from the administration's position.[90] Then Senator John F. Kennedy, who would make the handling of the crises an issue in his successful run for the presidency two years later, argued, "The weight of military, diplomatic, political, and historical judgment would dictate a contrary policy concerning Formosa."[91] Senator Hubert Humphrey suggested that the island dispute "be given to the War World Court to decide, and that the possibility of Formosa as an independent nation under the umbrella of United Nations guarantees and protection should be thoroughly explored."[92] Sensing the national mood, even Republican congressional leaders backed off unconditional support for defending the small islands. One China bloc senator commented that the U.S. position should be: "We will defend Formosa with everything we have but I do not believe it is in the national interest to go war over Jinmen and Mazu."[93]

Public opinion was foremost in the mind of Eisenhower; attitudes, while not pro-PRC, counseled against so hard a line as to lead to war. In 1957, a Gallup poll conducted June 6–11 asked, "Should the United States permit United States newspapermen to visit China, or not?" 57 percent answered in the affirmative; 26 percent opposed such a move. A poll of February 7–12 asked a sample of Americans the following question: "Premier Zhou Enlai of Communist China has offered to meet with United States Secretary of State Dulles to see if better relations can be worked out between the two countries. In your opinion, should Dulles meet with him or not?" 54 percent said yes, with only 30 percent agreeing with Dulles by being opposed to such a meeting. The same poll, however, found opposition to admitting Communist China to the United Nations as a member at 70 percent. In January 24–29, 1958, Gallup conducted a poll and found the level of opposition in the United States to China's admission to the United Nations at 66 percent. In specific relation to the second Taiwan Straits crisis, Gallup's poll conducted from September 10 to 15 asked of "those who said they had been following the crisis over Jinmen and Mazu: Would you like to see the United States work out a solution to this problem in the United Nations before we get more involved in a military way in the fight over these two islands?" 91 percent of the sample answered yes. 61 percent also agreed with the suggestion that "Formosa be neutralized—that is, put under the protection of the United Nations."

If the 1950s did not bring any great change for the better in Sino-American relations, other forces were at work to create the opportunity for eventual rapprochement. The second crisis was also another confirmation to the Chinese of the bitter lesson that they had learned in Korea and in the first straits battle: Russia was of limited utility as an ally. In both crises, Soviet "assistance" was largely rhetorical and restrained. Lucian Pye comments, "They [the Russians] understood, in spite of journalistic speculation, that there was no 'missile gap,' that Sputnik did not fundamentally alter the realities of power, and that the best they could accomplish was to bluff equality, not assert superiority."[94] Khrushchev, above all, did not seek to engage in brinksmanship. As a Russian historian has observed, "Khrushchev seemed less confident that he could push the West around with relative impunity; he seemed more sensitive to and wary of the danger of local wars escalating into a general war."[95] In fact, the Soviet premier sent a letter of protest to Washington in China's favor only the day *after* Zhou had renewed his call for ambassadorial talks. At the same time, Moscow's distaste for Mao and his brand of social upheavalism—from brinkmanship with America to forced collectivization and agriculture—was growing and becoming more open. The reciprocal feeling, in Mao and his leadership, was that China should develop its own strength independently of Russia's nuclear umbrella.[96]

Finally, 1959 was a fateful year within China. Mao faced a near rebellion in party ranks over the failures of his grand self-reliance scheme of the Great Leap Forward. Such madcap innovations as backyard steel mills, people's communes, and the extermination of sparrows brought famine and chaos that cost the lives of tens of millions.[97] The critical political moment was the Lushan Conference held from July to August; Mao again proved his true genius in party politics. Old comrade and Long March veteran Peng Dehuai was genuinely concerned that the meeting was not addressing the disaster China was facing but rather the absurd policies that were being promulgated. He wrote a private memorandum to Mao expressing his concerns—which unfortunately for him was then circulated among the leadership.

Actually Peng did not oppose Mao; the latter, however, accused Peng of conspiring with the Soviet Union against him.[98] Mao had indirect evidence to back up this accusation, since Peng had just visited Russia to a triumphant reception.[99] Mao also won the day with a sensational speech in which, bursting into tears, he admitted the failures of the people's commune system and bemoaned, "I deserve to have no offspring."[100] However, he exhorted the country to rally behind his leadership and to keep up the struggle against American imperialism. Peng was purged, spent sixteen years under house arrest, and died during the Cultural Revolution. Mao's star, however, was temporarily eclipsed as undisputed head of the party.

For the Russians, this situation confirmed China's tailspin into Neanderthalism. In late August, Moscow announced that it would no longer honor the defense technology agreement it had with China and withdrew all its military scientists from the PRC. In his memoirs, Soviet Deputy Foreign Minister for the Far East M. S. Kapitsa states that the casualness of Mao's reaction to Eisenhower's nuclear threat convinced the Soviet government that the Chinese were "too irresponsible for the possession of the ultimate weapon."[101] The Chinese, in turn, chafed at such an attitude from an ally. By 1956, the nuke threat repeated by Eisenhower and by the American press convinced Mao that China must have its own arsenal of mass destruction. Mao asserted, "If we are not to be bullied in the present day world, we cannot do without the [atomic] bomb."[102] There were many other points of tension as well: China was pricking India, a nominal neutral and a state the USSR was courting. In September, Khrushchev and Eisenhower met as part of the premier's two-week visit to America. Talks resulted in no treaties, but obviously the Cold War was warming in the direction of greater rapprochement between the superpowers.

Observing these events, Mao admitted, for the first time, that he was more impressed with the conduct of the Americans than that of the Russians.[103] Indeed, it was a few months earlier that Zhou made an important and (from the point of view of the historian and psychologist) revealing comment about his chief antagonist in the diplomacy of the 1950s: "Dulles says that he is hostile toward China. The United States will not do trade with but conduct a trade embargo against China. He is a chief reactionary but thinks and acts in one and the same way. His words are logical."[104] In short, Mao, like all paranoids, tended to trust his enemies more than his friends: hence his future faith in Nixon. On a personal level—at which Mao took everything—it must have been galling as well that Khrushchev received the grand tour while his own chief diplomat could not even get a handshake.

It would take more than a decade and yet more tension between America and China before a president with impeccable anti-Communist credentials would have tea with Mao in the Forbidden City. Jinmen and the offshore islets would feel the weight of PLA gunfire (a few shells every other evening) or razzias by "water ghosts" until 1979; even as late as 1995 the PRC would send missiles into the straits to express its disapproval with Taiwan. But as 1960 dawned, and another president came into office, a new era of the Sino-Soviet-American triangle was about to begin.

4. Pictures of the Crises

With the end of the Korean War, the long hard freeze in Sino-American relations had begun. It would last almost two decades. For Henry Luce and the Lucepress, including Time *magazine, this state allowed for the full expression of Manichean China policy. The Chinese government was unrelentingly evil, its leaders were beasts, and almost all its projects led to human suffering and disaster. China was either the Soviet Union's puppet or its partner. Chiang Kai-shek, a long with the citadel of Taiwan, was the great defender of the West.*

Several trends marked China attitudes in Time *in the 1950s. First, Luce's thirst for reconquering the mainland led him into editorial trouble. In the on-again/off-again shelling between the Nationalists and the Communists over tiny offshore islands, Luce saw an opportunity for war and the liberation of China. At one point, the magazine even advocated nuclear strikes against the mainland. The Eisenhower administration also made such considerations. But Luce and the General, both practical men, soon realized that they had gone far beyond the limits of American popular opinion. The American people, presumably among them* Time *readers, were anti-Communists and supported the survival of the Nationalist regime in Taiwan, but they had no taste for yet another land war in Asia, especially when the issue at hand was fingers of land off the coast of China that were no bigger than a city park.*

Sporadic diplomatic negotiations with China in Paris, Geneva, and then Poland led nowhere. At the same time, almost all American journalists were frozen out of China, so Luce relied on defectors, rumors, Nationalist leaks, and speculations about the true state of events in China. For images, ironically, he often had to use those provided by the Chinese government—but these, as we shall see, were stuck by Time *magazine with oppositional captioning.*

An overall survey of the China images in *Time* of 1954 to 1959 finds them, not surprisingly, peaking during the heights of the crises. *Time* had always paid some attention to China, but the rain of fire on the offshore islets gained the newsmagazine's attention as much as it did President Eisenhower's. In terms of what was going on in the images, for each year the main focus was generally on international relations and policies or military affairs and fighting. The frequency of images per year as well as the percent of the total number of pictures of the entire period of the visual analysis, 1949–1973, are shown below in Table 3.1. Other ways of measuring the amount of coverage follow: table 3.2 tabulates the number of cover images devoted to China; table 3.3 describes the total surface area of China images by year.

Table 3.1. China Pictures by Year, 1954–1959

Year	*Freq.*	*Percent*
1954	46	2.5
1955	47	2.5
1956	26	1.4
1957	30	1.6
1958	36	1.9
1959	43	2.3

Table 3.2. *Time* China Covers by Year, 1954–1959

Year	*Covers*
1954	1
1955	2
1956	1
1957	0
1958	1
1959	2

Another standard measure of the coding of these images was "social dysfunction," an evaluation of negative events going on in the country as a percent of all the pictures concerning the nation. Here, too, the numbers match events. When China is bellicose, *Time* reports bad things going on in China. Table 3.4 breaks down dysfunctionality by year. Table 3.5 looks at the main kinds of dysfunction.

The exact contents of the images, and the editorial intent of the whole picture cluster of words and images, is, as always, more revealing. To begin, 1954, in the wake of the Korean War, was a heavy year for China-related imagery. *Time's* January 11 issue begins with a People section society photo of Generalissimo Chiang and the Missimo arm in arm and smiling with the chairman of the Joint Chiefs of Staff, his wife, and the assistant secretary of state for Far Eastern Affairs. The description in the text of the photograph relates, "The visitors struck a family-album sort of pose." No line probably better summed up *Time's* attitude toward the Chiangs after the Korean War's solidification of its hope for and support of Nationalist China on Taiwan. The family album is followed up in the

Table 3.3. Surface Area of China Pictures by Year, 1954–1959*

Year	*Freq.*	*Percent*
1954	1,277	2.8
1955	1,722	3.9
1956	622	1.4
1957	1,052	2.3
1958	706	1.6
1959	1,109	2.4

*This table shows the yearly surface area of China pictures in *Time*. The calculation of surface area is based on:
Two pages = 187 square inches.
One page = 93.5 square inches.
Half page = 46.75 square inches.
Quarter page = 23.38 (rounded) square inches.
One-ninth page = 10.39 (rounded) square inches.

Table 3.4. Social Dysfunction in the PRC by Year, 1954–1959

Year	*Percent*
1954	54.3
1955	57.4
1956	73.1
1957	30.0
1958	63.9
1959	58.1

January 18 issue with a photograph of Vice President Richard Nixon and his wife visiting an orphans' home in Formosa during a state visit. The "American politician's wife with adorable Chinese tot" genre picture would become a staple of China coverage, and indeed would repeat itself when Nixon visited the PRC in 1972.

Here, the accompanying story tells us, appealing to a smaller Chinese mass, "Since Nixon's return, and partly as a result of his findings, certain viewpoints are gaining headway in Washington. Among them . . . the U.S. must not even talk about recognizing Red China. One reason is such talk would discourage thirteen million overseas Chinese in Southeast Asia from their present strong anti-Communist drift." As we have learned, among *Time's* code phrases for its support of a policy position is to say that it is "gaining currency" or "headway" among experts or in Washington. The battle for the hearts and minds of the diaspora was also vital. A map in the same issue shows the "lengthening chain" of states allied in the containment of Communism. Russia and China are shown as one continuous red blob in the Northern Hemisphere. It does not require too much speculation to appreciate how such images must have looked to the America watchers of the PRC: a "family-album sort of pose" for Henry Luce was a *causus belli* for Mao.

A series of pictures in the February 1 issue of *Time* shows "anti-Communist Chinese prisoners" captured in Korea who are being released from internment in neutral India to be transferred to Formosa. The "march of free men" is lavished

Table 3.5. Presence of Most Frequent Social Dysfunction by Year, 1954–1959

Year	*Dysfunction*
1954	Political/governmental struggle
1955	War
1956	Persecution of dissidents
1957	War
1958	Political/governmental struggle
1959	Persecution of dissidents

with a two-page spread. In contrast, the March 1 *Time* magazine has a feature story on railroad building in China; the accompanying map shows the "Red links." The story is full of descriptions of the length of the line and the number of workers involved in railroad production. At the same time, a subitem speculates on whether Mao Zedong is ill. Another article, including a photographic portrait, describes Chen Cheng Chiang Kai-shek's designated heir the presidency of Taiwan. It is almost wholly laudatory, including such descriptions as: "Not even his foes ever accused him [Chen] of the slightest personal corruption." The March 8 issue of *Time* is also focused on Formosa, with a commentary drawn from an interview with "China's most honored scholar, Dr. Hu Shih." Hu critiques the idea of a third way in the fight between Communism and anti-Communism. This point fits into *Time*'s advocacy program because, in the "linked chain" against Communism, countries like India that had talked of a "third way" (i.e., to be neutral in the Cold War, a position *Time* detested) are shown as blank spots. We may read this as an allusion to those nations' alleged empty-headedness.

A story on March 15 about agriculture in Red China details some of the Communist techniques of socialization for a general critique. *Time*'s lesson for its readers is summed up by the opening line: "The Communist masters of China are having trouble with the peasants." Interestingly, a propaganda leaflet is shown as an illustration to "the problems." However, it is not a PRC leaflet but a Nationalist one of the kind dropped onto the Chinese mainland. It displays Chinese soldiers appropriating farm goods, including the family pig, outside a hovel, while peasants holding sticks rise up to smite their persecutors. The accompanying story does mention that this is propaganda, but argues that it is based on the actual fact of peasant discontentment. Yet Chiang's regime does not receive unalloyed praise in *Time* magazine. The March 22 issue features a photo-portrait of a former diplomat and an item about his resignation from the Nationalist government. His charges, including corruption and incompetence, ring familiar to those held of the Nationalists during World War II and the civil war. The story, however, ends with a counterattack on Wu by Nationalist officials. In short, *Time* does not whitewash Formosa, but neither does it call for radical change.

China is featured in the news again in *Time*'s May 3 issue, with a photograph of Zhou Enlai standing smiling next to a North Korean leader. The photo is credited to the United Press, and it is a handsome picture, with both men nattily dressed and jocular. The accompanying story on the imminent French defeat at the battle of Dien Bien Phu in Indochina cites "the size and effectiveness of Communist China's professional help" as one of the key factors grinding down the French. The photo's subcaption sums up *Time*'s attitude toward the smiling diplomats, "with blood and pious talk of peace."

The May 10 issue of *Time* magazine continues this theme and features as its cover story "Red China's Zhou Enlai," who is "waging war and talking peace." The painting shows a smiling Zhou in a handsome portrait that the subject, although he probably did not pose for it, would no doubt have been quite happy with. But *Time's* editorial stance is not-so-subtly employed. Zhou is the friendly face of Red China, *Time* seems to be saying, but in the background we see the *real* China (see figure 3.1).

Behind bamboo bars is a frightening, scaly dragon, complete with long horns, hypnotic eyes, flaring nostrils, and claws that grab the bamboo, seemingly about to wrench open the bars to escape and charge—toward the onlooker. This is no "mysterious" dragon or "friendly" dragon or even "exotic" dragon—it is a beast that could be cast in a science fiction horror movie or in a tale from Tolkien. The message is clear: the terrifying creature is the true China, while Zhou is but a smiling mask. Several of the pictures inside visually strike home at the terrors of the dragon. There is an uncredited photograph of a large number of Chinese people with their heads bowed, witnessed by a crowd, many of whom wear the traditional sampan hat. The caption and subcaptions read, "Communist land reform: Landlords about to be shot . . . by meticulous use of murder, terror, force, and persuasion." The accompanying story enumerates the charge that "never has centralized authority been more fierce, continuous or complete in China." As we shall see in coverage of China after 1972, the portrayal of such terrors was largely to fade from American newsmagazines with the prospect of China as ally and as trading partner.

Zhou's diplomatic sojourns are featured in a number of other pictures throughout the year, all with either negative captioning or pejorative description in the accompanying stories. The Chinese dragon reappears in cartoon form as a squat, armored, komodo-like creature in the July 19 issue of *Time* (see figure 3.2).

The accompanying story discusses the controversy over whether to seat China in the United Nations. India and Great Britain are cited as supporting "peaceful coexistence," but the U.S. Congress is described as almost uniformly hostile. Lyndon Johnson, then Senate minority leader, is quoted as saying, "The American people will refuse to support the United Nations if Red China becomes a member." The cartoon is taken from the *Manchester Guardian.* This is another technique of *Time* magazine—when it wants to criticize another country's actions, it uses a political cartoon from that country. The dragon, labeled "Chinese Communism," looks on as caricatures of the English prime minister and India's Nehru read a document titled, "The lions lie [*sic*] down with the lamb." Meanwhile, an especially plush lamb looks on with sly, knowing eyes; he is labeled "W[estern] Democracy" and conceals a machine gun. In short, *Time* seems to be using this cartoon to express not only the duplicity and evil intentions of China, and the deplorable inaction of nations like Great Britain and

Figure 3.1. *May 10, 1954:* Time *cover painting, "Red China's Zhou Enlai." (Artist: Boris Chaliapin. Copyright 1954 by Time, Inc. Reprinted with permission.)*

Figure 3.2. July 19, 1954: Cartoon, "Peaceful Coexistence." (David Low, reprinted from Manchester Guardian. *Reprinted by permission from Solo Syndication, London/Associated Newspapers.)*

India, but also that the true spirit of Western democracy is to be "good" while remaining ready for the eventuality of treachery. It is a rejection of the idea of Nehruesque peaceful coexistence in favor of armed coexistence.

Time concerns itself in the next several issues with the collapse of the French position in Indochina and often makes the contrast between the peaceful words of the Communists and their warlike actions. China also morphs, in a map of August 2, into yet another monster, this time a Chinese octopus complete with slanted eyes shown with its tentacles wrapped around or reaching for almost all of Southeast Asia. This theme is continued in the weeks to follow. A cartoon shows a hospital ward in which the patients are all Southeast Asian countries. A surgeon, labeled "Chou," wheels in an empty bed; the caption asks, "Who's next?" Later, Burma is featured as another Communist Chinese target. Then, in the issues of September 13 and 20, British prime minister Clement Atlee, of the Labor Party, is featured for his trip to China. The story and the picture captions are as contrarian and disdainful as one could imagine. For example, a description of Atlee's rhetoric is given as coming across "a little oddly over the sound of Communist artillery hammering Jinmen and Mazu and the howls of Red Chinese leaders for the . . . liberation of Formosa." Later stories focus on the defense of Jinmen and Mazu and the fifth anniversary of the Communist victory in China. The latter photographs, incidentally, are shown without much internal negative captioning, but the overall descriptions of the picture series (images such as parading tanks, female athletes, and marchers) "offer striking evidence that the Moscow pattern of regimentation has taken hold in the ancient Eastern capital." *Time* completes its coverage of China in 1954 with a series of stories and illustrations about Chinese expansionism aimed toward the rest of Southeast Asia.

Time begins 1955 with a set of world maps in its January 3 issue that detail the "watchman's rounds, 1954." For a number of countries around the world, diplomatic status and treaties or agreements related to the Cold War are cited. Then on January 31, after a number of issues with no China imagery, *Time* focuses, as it would for years to come, on Formosa. The internal story details the seizure by the "Reds" of a small island 250 miles northwest of Formosa, and indicates that the next step in Communist planning was to attack another set of tiny offshore islands, the Tachens. The diplomatic response of the Eisenhower administration is carefully critiqued. On the one hand, several of the small islands are listed as being militarily indefensible; but then, Eisenhower and Secretary of State Dulles are reported as being caught by a reporter's question about whether they favor a "cease-fire." *Time*'s opinion is made clear when the story argues, "The U.S.'s calculated risk in the field of propaganda had misfired. In Asia, where the shooting is in progress, the burning interest is not in who is for peace, but who is" the strongest. Again, *Time* tends to speak of "Asia"—its billions and many nations—as a single entity. The story continues, "A cease-fire would leave the Chinese Nationalists with no hope for the future. Talk alone had done at least part of the damage that a cease-fire would do. It had demoralized U.S. allies in Asia."

The pictures illustrate *Time*'s bellicosity. Chiang Kai-shek, in his characteristic cape, is shown arm in arm and beaming with the vice admiral in command of the U.S. Seventh Fleet. The subcaption reads, "Asia wondered who is going to win." Another story discusses the fate of a number of Americans still held as prisoners in China. A picture that was provided by the Chinese government to visiting UN Secretary-General Dag Hammarskjold is shown: it depicts some Caucasian men feasting and smiling at a banquet table; the subcaption warns us that they and their families are "torn between doubts and desire." Inside, a map showing most of Formosa and the adjoining Chinese mainland enumerates some of the military actions of the past weeks. Importantly, the "defense line guaranteed by the U.S." covers the Pescadores Islands nearer to Formosa. The islands of Jinmen and Mazu, just off the mainland Chinese coast, are not shown between the lines. As the story relates, "Now the Chinese Nationalists fear that, if the impression of U.S. feebleness and indecision spreads in Asia, the Red siren calls will be so much the stronger—and will fall on ever more receptive ears." This is another allusion to the battle for the hearts and minds of diaspora Chinese.

February 7's *Time* follows up with a cover story devoted to "Defense of Formosa: U.S. Strategy in Asia." Admiral Pride of the Seventh Fleet is shown in photographic close-up, presumably aboard ship, overlooking a seascape. *Time*'s editorial views are enforced again when, while in the previous week it fretted about uncertainty and weakness in the U.S. administration's position on Formosa, now, as *Time* relates, "Ike stopped the rot" through "strong leadership." The newsmagazine was referring to a congressional resolution (passed 409 to 3 in the House and 85 to 3 in the Senate) that supported Eisenhower's call to "'clearly and publicly establish the authority of the president' to use U.S. forces as he deems necessary for the protection of Formosa and the Pescadores [with] authority . . . broad enough to cover defense of 'related territories.'" One picture in the issue shows Eisenhower signing the Formosa resolution, with the subcaption, "Strong leadership brought united determination"; another image shows Chiang and Pride on a U.S. aircraft carrier.

Time's coverage of the Jinmen and Mazu issue illustrates an important reciprocity effect for the "intrinsically negligible" islands. The islands are not technically part of Formosa and never were. They are only a few miles off the coast of mainland China. They are defensible only by extraordinary means. In short, they are important as a symbol only because, by cooperation, the United States, the Nationalists, Henry Luce, the China bloc, and, ironically, Communist China had *made* them into a symbol. Who was served by this process and procedure? In their diplomacy with the United States, the Chinese often were surprised at how Americans seemed to place great weight on things that were, to the Chinese, insignificant. That a few U.S. citizens were held hostage in China, for example, was considered to be a major concern by the Eisenhower White House; it never would have been in the reverse case. One could argue the following: Luce and the China bloc beat the drums for Jinmen and Mazu because they wanted an issue to confront China with after the "lost opportunity" of Nationalist participation in the Korean War. So the trivial islands, which could not be justified as strategic or economic fulcrums, became instead vital metaphors for resistance to Communism.

But a symbol is, of course, a constructed system of denotation and connotation. The great iconologist and mythologist Philip Wheelwright noted that what most kinds of constructed symbols "have in common is the *property of being more in intention than they are in existence*."[105] For example, men in battle do not die for a piece of cloth; they die for their *flag*. In the stories throughout the 1950s concerning Jinmen and Mazu, *Time* repeatedly hit one note: their surrender would dishearten the anti-Communist cause throughout Asia. In support of such a proposition, *Time* would deploy its own arguments, but also often quote from a local—typically Hong Kong—anti-Communist newspaper, or an "ordinary" Chinese person such as a fisherman or businessman. Conversely, Jinmen and Mazu were useful as symbols to Communist China. The PRC could not seize Formosa; any attempted military venture across the breadth of the straits would end in humiliation and defeat. Escalating tensions by threatening and bombarding coastal islands made them appear, to themselves and to their own people, as if they were actively battling the "running dogs of American imperialism." The upside was that other parties, such as Europeans, might try to mediate, with resulting diplomatic gains for China. A confrontation with the United States over Jinmen and Mazu was thus considered to be a net plus both by Communist China and by its enemies. In that sense, again, opposites worked toward opposite goals via the same route: Chiang and Mao both needed the islets for their own purposes. So, I argue, did Luce. It was not quite a *Wag the Dog* scenario in which a war was created from scratch (and in the photo lab) to distract the public, but for almost all parties involved it was the functional equivalent of a pseudobattle or, to paraphrase Eisenhower, a Gilbert and Sullivan war with real blood.

These themes are continued in the April 4 issue of *Time*, where three maps show detailed views of the "islands of war" of Jinmen and Mazu. The accompanying story asks, "Will the U.S. soon be at war again?" The story notes that the islands are defended by up to sixty thousand Nationalist troops, but that they could not survive without U.S. protection. Eisenhower's dilemma, as the story describes, is that "If he decides to use atomic weapons, which the military now regards as standard pieces in the modern weapons arsenal . . . [ignorant objectors] will predict that a hydrogen holocaust will inevitably follow. [Yet] despite Red China's fanatical aggressiveness, the best U.S. intelligence is that the Communists are not ready for World War III." The modern reader will be startled by such words—and they were hardly more plausible in 1954. But even a reader from the 1950s might ask whether "the best U.S. intelligence" about China had proven as astute and prophetic about China's actions in the last few years as *Time* claims. (Indeed, only a few years ago *Time* was castigating the failures of "the best U.S. intelligence" about China's actions). In any case, the story continues, "If the U.S. bows before Communist aggression again, the watching Asian millions may finally decide that the U.S. is, as Red China has charged, merely a paper tiger, of no value as a friend and ally." Again, *Time* speaks for Asia's millions and assumes that Asia's millions speak with one voice.

Time's April 18 issue is devoted to the Formosa crisis. Chiang Kai-shek appears on the cover, white-haired, slightly more wrinkled, but calm in five-star generalissimo uniform and exuding a visionary demeanor. Behind him is a shoreline, complete with defense works, a bunker, and a soldier standing guard. The horizon glows red, and in the midst of the mass are the large star and four small star symbols of the Communist Chinese flag. The image is full of intriguing connotations. Is this a beach at Jinmen or Mazu, or is it Formosa that is under siege? Since we assume that this is an east-facing beach, Chiang, by looking slightly to his left, is gazing over his island but also across the Pacific toward America. Is he awaiting help from the United States? His serenity is marked; he is not screaming threats or panicking. He is, geologically speaking, much like the giant rock that is Formosa. Within the story, we are reminded that Chiang's name means "firm rock," and that "in a time of confusion, [he has] an air of tranquility." And what of the glowing PRC? Is this a firestorm of Communist aggression, a creeping "Red tide" of Maoist disease, a polluted atmosphere, or a nuclear explosion? Inside the magazine, Chiang is featured in several photographs: on a rowboat on a lake with Madame Chiang and playing with his grandsons, who, dressed in jeans and flannel shirts, look exactly like American schoolboys playing cops and robbers. Other Nationalist leaders are featured, along with maps of the Taiwan Straits region and "Red China."

Pictures in *Time* for the rest of the year 1955 focus on diplomacy around the Jinmen and Mazu issue; American forces in the region; and a relatively neutral story-and-picture essay of "Red China's capital." One photograph in the October 3 issue, however, shows some rare direct imagery of Communist oppression: Roman Catholic bishop Alfonso Ferroni, a skeletal figure, is shown seated talking to a "fellow priest in Hong Kong." With faltering words he divulges, and we learn from a physical examination, the forms of torture that he had suffered at the hands of the Communists. The story also describes a complete crackdown on religious activity in the country.

After a hiatus of several months, *Time* resumes its coverage of the People's Republic of China on March 5, 1956, with a bang and a whimper. A painting on the cover shows a smiling, if fiendish-looking, Chinese man in a four-star general's uniform, introduced as "Red China's police boss Lo Jui-Ching." *Time*'s editorial backdrop is a gigantic bloody handprint. Inside, the issue details the "high tide of terror" in Communist China. Uncredited photographs show the mass executions of "peasant landlords." The story details a holocaust going on in China that includes up to twenty million "deprived of existence, done away with, or otherwise disposed of." *Time* raises the problem of the inability to "see" such evil, saying that "because it is hard for the mind to visualize so vast a slaughter in human terms, the Communists have been able to reap an advantage from the very size of their funeral pyre: many Westerners, finding the monstrous incredible, cannot see the blood on the hand of pretended friendship proffered by Chinese Communist leader Mao Tsetung." In profiling the terror, the magazine makes frequent references are to the Soviet precedent.

The dilemma raised by *Time* is a real one. A few critics in 1972 would note that Richard Nixon would shake that very hand with blood on it—and there is no statute of limitations on mass murder or crimes against humanity of which Mao was surely guilty in 1956 and in 1972. On a larger scale, the greatest horrors of the world were those perpetrated largely off camera. The slaughter of native Formosans that took place shortly after the Nationalist takeover of the island received minor and mostly written description in the American press. The visual documentation of Hitler's holocaust only became extensive after the defeat of his regime, likewise for the genocide in Cambodia by the Khmer Rouge and the victims exterminated in Stalin's gulags. China's crimes against its own people were almost always out of sight for the period in which foreign journalists could not travel to China; but then when China "opened up," the foreign press could not cover most of the dark side of China's rule, especially in Tibet.[106]

One minor item, appearing in *Time*'s May 14 issue, was a picture credited to Jean Renoir (for *Life* magazine) of two slender Chinese women strolling down a sidewalk in fashionable qipao dresses and heels. The accompanying story mentions a trend of de-emphasizing the plain "liberation uniform." *Time*'s editorial stance does not fade even in this fashion piece, however: the story concludes, "Chairman Mao and his gang appeared in their old dark suits, more like a phalanx of rigid revolutionaries than flowers in bloom."

Time's News in Pictures section of July 2 shows photographs taken by a German photographer who was granted a visa into China. One of his studies was of some of the "sixteen remaining U.S. prisoners of war who chose to stay in Red China after the Korean truce." They are described as "turncoats," and the picture captions, as usual, work to undermine the relatively benign content of the "Red" images: a black soldier, seen dancing with a Chinese woman, had, we are told, "circulated anti-U.S. petitions in prison camps." Bicycle riders "are free to move about the capital but are watched by the Red government." Another prisoner having lunch is a "wartime collaborator with Red guards in PW [*sic*] camp." Two other Americans huddled around a stove as they "study Chinese language" are "willing students of Communist propaganda charged with trying to influence other U.S. prisoners during war." These images, of course, can be contrasted with the coverage of the more numerous Chinese prisoners who chose to defect to Nationalist China, who were portrayed as seekers of freedom (and builders of a homespun Statue of Liberty). As historian Adam J. Zweiback notes in his study of the case of the "turncoats" (actually numbering twenty-three in all), "Since the vast majority of American POWs did not go over to the Communists, many newspaper and magazine articles looked for quirks in the backgrounds of the 'turncoats' that made them particularly susceptible to Chinese mind-control techniques."[107] According to *Time*, the alleged explanation for one prisoner was that he was raised in a "slum."

Another case of contrarian captioning comes from the July 23 issue of *Time*. There is a report about the antivermin campaign in China, Mao's disastrous idea—although treated humorously here—that, since small birds eat grain, it was the duty of the Chinese people to exterminate these pests. The campaign was successful: people killed millions of sparrows. The result was a huge increase in the insect population, with a great negative impact on Chinese agriculture. But *Time* does not delve into these more serious ramifications, which would have been suggested to them if they had consulted any wildlife biologist, but rather takes the opportunity to poke fun at Communist statistical-mindedness. A "Chinese Communist children's poster" that, we are told, was captioned "Driving the Americans into the sea," is reprinted. Cherubic but angry and determined-looking Chinese tots are variously working at munitions factories, driving tractors, or wielding machine guns, manning battleship turrets and fighter planes in the guise of soldiers, sailors, and airmen. Two figures marked as "U.S.," one who looks MacArthuresque, cringe in bedraggled terror, showing their spiked teeth. Above it all floats a white dove of peace. It is an amazing image that is critiquable on aesthetic and political grounds. First, the figures are directly in the style of World War II–era Japanese anti-American propaganda posters. *Time* could have provided such a visual contrast and decisively embarrassed China's cartoon propagandists. Second, the image, viewed through the Western eye, can be read as an abomination: children operating heavy machinery, forced to labor in dangerous work, and employed as soldiers. But *Time* either thinks these aspects of the image are self-evident or does not recognize them. The subcaption's contrarian sentiment contrasts the image to: "On the home front, 2,832,000 dead sparrows and 869,525 rats."

In the weeks to follow, the few China pictures focus on refugee problems, guerrilla warfare in Burma, and dam projects within the PRC. The contrarian captioning continues in *Time*'s October 1 issue: an Eastfoto of "Red China's Mao Tsetung" at a meeting illustrates a story about the Chinese Communist Party boasting of its great industrial achievements, in effect mass totalitarian oppression. The subcaption of the photo reads, "Beneath the boasts, uncomfortable truths." The story does not end in complete ridicule, however. It argues that the targets of China's five-year plan, "high as they are, may well be achieved. China's Red masters were doing well—too well for the peace of mind of the free world." If it sounds problematic to call the Communist regime a failure one week and a success the next, the contradiction continues in *Time*'s December 10 issue, with a picture of a flower-bedecked Zhou Enlai, arm in arm with India's Nehru, and captioned, "Cunning, cool, and curious." Interestingly, the story notes that Zhou made several statements calling for better relations with the United States. *Time* does not dismiss these but leaves an open question of what the U.S. administration's response will be.

Time's January 7, 1957, issue shows a still from the CBS Edward R. Murrow news program *See It Now*. The caption and subcaption read, "Communist

Zhou Enlai: Less than his press." The accompanying story within the television and radio criticism section calls him "by reputation, one of democracy's most eloquent enemies . . . [who] failed in appearance and performance to live up to his reputation." Zhou is the featured butt of *Time*'s captioning in the issue of January 21: he is shown in a fur coat, arm in arm with two Polish Communist leaders. The accompanying story relates that he toasted them "based on principles of equality and mutual respect." The caption, however, tells us that we are seeing "Red China's Zhou Enlai between 'equals'; little brothers are subservient to Big Brother."

After a long China picture drought, *Time*'s May 6 issue contains a large section on "masterpieces of Chinese art." The featured pieces are vases, calligraphy scrolls, paintings, porcelain, and bronzeware; although none of the seventeen photographs of the art contain modern political content, and indeed every piece is of dynastic origin, all come from Formosa, where the accompanying story tells us the Nationalists have "complete[d] the epic task of saving from Communist hands the art treasures assembled over the centuries and collected in the Beijing Palace Museum and Nanjing Central Museum." The story is ostensibly a pro-Nationalist piece, and fits into *Time*'s inclination to glorify the cultural heritage of China before the Communist takeover.

Time's May 27 issue shows a picture of a dour, rumpled Mao holding a cigarette. The subcaption reads, "Let all the flowers bloom, tra la tra la." The reference is to the "hundred flowers" campaign then begun in China, where for a short period intellectual dissent was encouraged, only to be persecuted later. *Time*, in hope, reports that the campaign was part of a denial of "fundamental propositions of Soviet ideology." As for the particulars of the flowers blooming, *Time* reports them, but its editorial tone is not trusting that the result will be good. In this, *Time* is absolutely correct. The campaign would unleash so much criticism of the CCP that Mao would launch an "antirightist" campaign in response.

Time's June 3 issue shows a picture from a rare flare-up of anger between the United States and Formosa. A U.S. Army master sergeant stationed on the island was being court-martialed for the killing of a Chinese man whom, he said, he had "caught spying on his wife." The story details anti-American riots and sentiment throughout the country. The picture shows the widow of the Chinese man in front of a protest sign outside the U.S. embassy. It is subcaptioned, "At her heels, a frenzied mob." *Time*'s June 24 issue shows a picture of a young Chiang Kai-shek; it illustrates the review of Chiang's book, *Soviet Russia in China.* The review itself is positive, extolling Chiang's thesis that "peaceful coexistence" with the Communists is what led his regime to defeat, since the Communists were able to destroy China from within. A footnote informs the reader that "a two-part condensation is being published in *Life* this week and next." *Time* does not mention that *Life* is a sister publication. Again, *Time* is not covering up the problems on Taiwan, just contextualizing them in Republican China's favor.

In the July 8 *Time*, yet another contrarily captioned picture of Zhou appears. This time it illustrates the crackdown on the hundred flowers campaign; the subcaption of the photo reads, "Instead of flowers, weeds." *Time*'s July 22 issue features a rare story on the problem of reporting on China from the West. A photo-portrait shows a Canadian reporter who entered China. Of course, the findings that *Time* reports are largely negative: the Chinese are poor, subject to brainwashing, engaged in a program of "free abortions and sterilization," and suffering an industrial slowdown. Nevertheless, the reporter does note that he was able to get several frank answers from people he talked to, and encountered few censorship difficulties.

Zhou is pictured with another caption that subverts the intended meaning of the picture creators in *Time*'s October 21 issue: he sits somewhat stiffly among some chanting and clapping American college students whom the accompanying People section story describes as "innocents abroad—part of [a group of] students who thumbed their passports at the State Department and a-walled* off to China last summer [and] got together with that jolly old minstrel Zhou Enlai for a clap-hands songfest." The subcaption is equally mocking: "Do re mi Chou." *Time*'s December 2 issue shows a smiling Mao standing next to an equally grinning Khrushchev in Moscow. The subcaption reads, "The flowers were blighted." The story is about a Communist meeting in Moscow to celebrate the fortieth anniversary of the Bolshevik Revolution. It emphasizes that Russia still demands that other Communist states recognize it as the capital of Communism and the leader of the Communist Bloc.

Time's China coverage of 1958, the year of the most acute crisis over the islands of Jinmen and Mazu between the United States and China, begins on January 6 with a photograph of Khrushchev and Mao standing and looking at each other with wide smiles. The picture is an illustration from Khrushchev's world tour, highlighting the rise of a "new class" of Soviet party leaders in the post-Stalin era. Interestingly, at this time of growing tension between China and the Soviet Union, *Time* magazine—which for many years previously had speculated on, even intended to provoke, such discord—was now proclaiming, "China's Mao was so convinced of the rightness of Khrushchev's policy reversal that he led the way for adoption of Khrushchev's manifesto." Later, on August 18, *Time* shows another picture of Khrushchev and Mao meeting, this time in Beijing. The story describes an even stronger crackdown in China on dissidents, and maintains that Mao had pushed Khrushchev to break off a meeting with Eisenhower. The story notes, "In the classic fashion of young dictatorships, Red China must rely on 'the threat from abroad' as a prop to internal discipline." Another story in the same issue features a picture and discussion of the schism of the

*A play on Great Wall and AWOL.

Chinese Catholic Church from Rome. It adds more details of religious persecution in China.

Time's January 20 issue features a small story on a Chinese "racket" of illegal immigration to the United States. The photograph accompanying shows the family of a San Francisco patriarch who allegedly ran an immigration scam to import almost the entire population of his home village in Canton. The story shows a certain ambiguity about dealing with the Chinese in America. A footnote calls the Exclusion Acts of 1882 and later "notorious." On the other hand, the piece does portray the immigration scheme as a great scandal and reports that the "elders of the . . . clan [in question] sensed the crisis, met in a laundry [and] decided to parade into the" U.S. Attorney's office "and ask for mercy." *Time's* editorial point of view, however, evinces itself in its assuring the reader that the greater Chinese community in Chinatown is attacking the scheme and that the instigator is "the most hated man" in the community. There is no evidence of any kind offered to prove this assertion. The story seems to be designed to make the case that this was an aberration, so we could argue that its editorial stance was benevolent toward the American Chinese community. Yet not once in the story is there a discussion of the utility of Chinese escaping from Communist China. One would think that this would be an opening for *Time* to make this point. The reader is left assuming that *Time* believes that Chinese have the right to escape from Communist China but only to Hong Kong and Formosa. The admission is made, however, that "Deporting them would be impractical."

Time's issue of February 10 continues its focus on the culture and glories of dynastic China. A series of bronzes from a traveling exhibit in a Washington museum are shown. There is no contrast made with the present state of Chinese culture (i.e., in Red China), but the implication is clear: once China had a glorious civilization; now it does not. For more than five months thereafter only a few minor China images appear in *Time* magazine, with some brief stories on Taiwan. One very attractive picture in the July 25 issue is of Madame Chiang on a U.S. tour giving a speech behind a flower-bedecked podium. Her comments about the struggle of freedom against tyranny and the worldwide battle against Communist imperialism are quoted uncritically. Her poignant visit to the dying Major General Claire Chennault is also described. It is about as favorable a story as is possible in a news magazine. In the August 4 issue, a heroic picture of Chennault behind one of his Flying Tiger airplanes accompanies his obituary.

The Jinmen crisis coverage begins in *Time's* September 8 issue. The heavy shelling of Jinmen and Chinese threats against it are detailed in the story; a map shows the "target islands" as well as the Chinese province of Fujian and Formosa. Interestingly, the U.S. Seventh Fleet, complete with American flag, is depicted as a white arrow interceding between the two. But there is another detail of interest in the map that seems puzzling to the modern observer. Formosa is listed as having six hundred thousand troops while Red China has "three hundred thousand troops" in Fujian. (This is twice *Time's* previous estimate of Nationalist manpower.) The story's ambiguity about Jinmen comes through in the notation that for China, the islands "were not worth anything like the ammunition expended," but to the Nationalists, these are "largely symbolic specks." In sum, *Time's* attitude toward this initial second Taiwan Straits crisis is much less bellicose than was its coverage of the first crisis. There is hardly a rallying cry to war, and more straight reportage. For example, a photograph in the September 15 issue shows "Jinmen civilians filling sandbags." The ROC has done "a beautiful job of preparing Jinmen against assault." The problems of supplying the island are also detailed: that is, Chiang needs American help to save the brave civilians.

A map in *Time's* September 22 issue shows an extreme close-up of Jinmen and the surrounding Chinese coastline. The accompanying story details the bombardment and resupply attempts of the island. Its main focus is on the United States escorting of convoys up to the three-mile limit. Several unsuccessful attempts to land supplies are described. The story concludes that the basic facts are "that the U.S. and Nationalist China could not assure the supply of beleaguered Jinmen without massive aerial bombardment of Red artillery positions on the Chinese mainland [but] the Chinese Communists could not hope to capture Jinmen by direct assault in the teeth of the U.S. Seventh Fleet." Then the story produces a dilemma for itself. On the one hand, it critiques the PRC for refusing to accept that its bombardment alone would reduce the island. On the other hand, the diplomatic problems of the United States are listed: lukewarm to no support from almost all its allies. *Time* does not go so far, however, as to call the American position unrealistic or "a refusal to recognize the facts." In this and other images from the crisis to come, *Time* takes a more neutral stance than it had previously.

In the October 6 issue, however, *Time* becomes bolder. A story details the increasing calls by Chiang Kai-shek to strike against the mainland. The linkage the story offers is that Chiang believes that there might be "a Hungarian-style uprising" that Chiang's forces might exploit. The story reports such prognostications without comment, but adds, "Chiang's impatience was a sharp reminder that the West could not shelve or solve Jinmen's problem simply by demanding that its defenders sit and take it . . . in a battle they could not afford to lose and were not allowed to win." Another story, focusing on the details of the struggle at Jinmen, notes, "To the weary, frustrated defenders . . . [no amount of supplies would be] a completely effective answer to the relentless fire from the unmolested Red guns on the mainland."

In *Time's* October 13 issue, in a section titled "Judgments and Prophecies: Mr. Dulles Meets the Press," a range of editorial comments on the crisis from different newspapers is displayed. The polarities are from strong criticism of Chiang Kai-shek to an attack against appeasement to a welcoming of a U.S. initiation of negotiations over the status of the islands. The four cartoons that are featured are all critical of Dulles: the basic theme seems to be that the United States is unable to do what it wishes to do in face of the "cold, hard facts." But in the November 24 issue, a Jinmen status story shows one of the first true atrocity pictures of *Time's* coverage of China. A photograph shows a woman, her face heavily bruised and bandaged, in a makeshift tent hospital. In *National Geographic* style, a baby nurses from her exposed breast (we note that *Time* would never have published a picture of a white woman doing the same). The caption reads, "Chinese victims of Red Chinese shells on Jinmen: where the senseless and the cruel become commonplace." The senselessness referred to is, as noted previously, the Chinese policy to shell Jinmen every other day. A countercriticism, however, may be that, while the People's Liberation Army is directly responsible for these people's tragic suffering, the United States and Nationalist China made the decision to keep women and babes-in-arms on the front line of a possible hot war with mainland China. These people are indeed victims, but it is unclear, morally, from whose cruelty they suffer.

Time concludes its major China coverage of 1958 with a bang. The December 1 issue features a cover of "Red China's Mao Zedong" (see figure 3.3).

Mao's porcine visage is adorned with pursed lips; his eyes look shiftily to the side in a mocking parody of the "visionary look" of great leaders. As a reference made to this contradiction, a banner atop the cover reads, "Let China sleep; when she awakens, the world will be sorry.—Napoleon." Behind Mao in the image, as if to emphasize the horrors of Red China, two bare-chested sweating coolies carry giant buckets tied to a wooden plank that is crushing their shoulders. Their expression, stance, and situation make them appear unquestionably like slave laborers. The interior description in *Time* states, "In the nine years since Mao Zedong took power in Beijing, China has become shrouded in a fog of ignorance almost as thick as in the days of Marco Polo." The story asserts, however, that, "determined to separate fact from fiction, *Time* correspondents in thirteen bureaus around the world began a mass assault on the bamboo curtain. Their chief weapons: interviews with scores of latter-day Marco Polos . . . .When the pieces were put together, what emerged was a documented picture of a national drive for power unparalleled in terms of human energy and forced sacrifice—but far less awesome in terms of results achieved, than Red China's masters (and nervous enemies) proclaim." *Time's* journalistic paradigm is intriguingly revealed in this "promise." The newsmagazine does not admit that it contributed to the fog of ignorance about China for several decades. It guarantees its readers gnostic knowledge that apparently even the U.S. government and others do not have. The pictures within the issue, however, are not necessarily negative: peasants working, Mao posing; there are no pictures of slave labor compared to that painted in the cover image. The story ends with an ambiguous and frightening

Figure 3.3. December 1, 1958: Time *cover painting, "Red China's Mao Zedong." (Artist: Bernard Safran. Copyright 1958 by Time, Inc. Reprinted with permission.)*

assertion: "That Mao has started China growing again is a fact of incalculable importance. If human beings can be reduced to mindless production-line cogs, Red China may one day achieve the stature for which its rulers yearn."

Time's January 5, 1959, issue contains another back-handed slap at Communist China and programmatic (as well as, in my view, problematic) praise for Chiang Kai-shek. The old Nationalist leader is described giving a speech in which, "in keeping with the constitution of 1947," he would not run for a third term as the island's president. Chiang is quoted as noting, "Close adherence to the constitution was 'one of our chief weapons in waging war against the Communists.'" The story does note in passing, however, "The G-mo himself intends to keep things in hand by retaining the powerful director-generalship of the Kuomintang Party." On February 9, *Time* takes another one of its excursions into an exhibit of dynastic Chinese art, this time painted scrolls from the Boston Museum of Fine Arts.

Time's March 30 issue contains an illustration for a story on "Mao's muscled minions." The photograph shows a PRC weightlifter at a competition in Moscow. The story notes how the Chinese athletes ascribe their dedication and medals to the inspiration of fighting "U.S. imperial aggression against our territory of Taiwan." Its reply is that "some Chinese claims bring only sniggers in the West." However, with the magazine's usual deference to sheer power, a Western observer of the Chinese team does note that, as *Time* puts it, "Mao's big brotherly encouragement to sport is no joke."

For the next several issues, *Time* devotes its China coverage to the flight of the Dalai Lama from occupied Tibet. *Time*'s gloating is obvious: for almost a decade, India and Prime Minister Nehru were the subject of opprobrium for appeasing the PRC and making self-righteous criticisms of U.S. anti-Communist policy. *Time* calls the incident the "most dramatic symbol of the Cold War's progress." The Dalai Lama, it says, "won a psychological victory. Red China's rape of Tibet stirred the neutralist powers of Asia as the Soviet rape of Hungary never had. With shock, Asians suddenly realized that there could be 'yellow colonialism' as well as 'white colonialism.'" Other publications are quoted supporting this line. *Time*'s point, historically, is correct: Nehru, who, within a year, would fight a war with China, had long refused to treat China's territorial ambitions as simply another form of colonialist expansion. But the sweeping generalization of "Asians" realizing any one thing is obviously *Time*'s hope rather than an accurate survey of all the several billion people of Asia's opinion on the subject. An interior map shows the "land in torment" of Tibet, along with the border areas from China to India. The Communist Party of India is described as being hotly criticized within the country. The story concludes with an allusion to the similar situation of Taiwan, quoting an Indian newspaper, "Let us hold our heads low. A small country on our border has paid the ultimate penalty for its temerity to aspire to independence."

Time's April 20 issue features a story on "the escape that rocked the Reds." A painting of the smiling but determined Dalai Lama graces the cover. Behind him is a peaceful lake overlooked by his monastery palace. Inside the magazine, pictures illustrate *Time*'s essential point in one caption, "Asians were beginning to see." Much hope is given to the revolt in Tibet as an "Asian Algeria," where one hundred thousand warriors prepare to fight for the holy monk. For *Time*, the Tibetan story, ignored for almost a decade, has now become a weighty counterpunch for its long-standing ire at neutralist and appeasement-minded Asians. The Dalai Lama, to confirm this editorial stance, is the subject of several weeks of photographic coverage, interspersed with a few negative stories about mainland China and positive articles about Taiwan. In these pieces, the focus is on internal troubles in China, and how border tensions between the Chinese and the Indians are heating up. China is reported to have made border incursions into India's Kashmir province: "India's new positive posture stiffens spines throughout southeast Asia." Perhaps now, *Time* is speculating, India will no longer be a blank spot along the ring of containment.

A cartoon in the September 14 issue taken from the *Washington Post* shows a pudgy, gap-toothed, warlock-visaged Mao standing on a barrel while stirring a set of boiling cauldrons labeled "India," "Laos," "Tibet," and "Bhutan." In front of this conflagration is an open-mouthed and open-handed Khrushchev quoted as saying, "We want to melt the ice." The reference is to his meetings with President Eisenhower. It is noted in the accompanying story that Moscow is ignoring rather than cheerleading for Chinese adventurism. The article concludes, "So long as Beijing continues to disturb the peace of the world, the West is unlikely to be very impressed by Moscow's mask of moderation." As a capstone to such a view, *Time* devotes its October 12 issue to "Red China's first decade: Work, purges, disillusionment." The cover picture is of "Red China's Liu Shao-Chi [Liu Shaoqi]." He is smoking a cigarette with an aristocratic holder and is shown almost completely in profile. He is deep in rumination, by implication over the troubles of China. The pictures within the periodical are mostly placid illustrations of street scenes and Liu and other Chinese leaders, but a diagram illustrates "the big fumble" of China's crumbling economic situation.

In *Time*'s October 19 issue, another map illustrates the border skirmishes of India and China. Again, the story concludes with an anti-Communist quote from an Indian newspaper criticizing Red China's "expansionism." Another cartoon, in the November 9 issue, ridicules Nehru's "tame neutralist policies [which are] blamed for much of the trouble [with China] and for India's unpreparedness to meet it." The cartoon shows Nehru, in his trademark jacket, with lines of worry and perplexity in his face, holding an olive branch and a nervous white dove. His opponent on a wrestling mat is a burly sumo wrestler. The cartoon is from the *Manchester Guardian* of England; *Time*'s captioning reads, "Opening round." A sign reads, "Communism versus democracy for the leadership of

Asia." In the background, a group of Asians dressed in various native styles cowers in fear. A Khrushchev figure looks on, hands covering mouth in confusion. The accompanying story ends on a buoyant note, saying, "At week's end, Nehru himself was beginning to talk tougher."

Time's November 30 issue focuses on ten years of Nationalist rule in Formosa. The story cheers, "For a decade's progress, a hearty how." The two pictures feature the Gimo and Madame Chiang and an aerial photograph of lush "rice paddies on Formosa." The contrast to China's "big fumble" is marked: Taiwan is a success; Red China is a failure. As a capstone to the eventful period, *Time*'s December 14 issue presents the Chinese dragon on the cover. India's Nehru is shown in his traditional costume, profiled, his hands to his lips, obviously in deep consternation and worry. In the background is a mountain range. A snaky serpent-dragon, eyes blazing at the reader, giant claw extended over and through a mountain pass, approaches. President Eisenhower's Asian tour is referred to as the "mission to unify friends and win neutrals." The story inside details a sea change among Asians from carping and criticism by Indians to the exhortation to "give the Chinese a bloody nose." *Time* points out that "India felt both angry and alone. The ruthlessness of Red China's behavior made a wreckage of some cherished convictions." Attacks on Indian border posts by the Chinese, and their "ruthlessness," are detailed. The editorial view expressed in the article is that the chastened Indians will now, it is hoped, admit the failure of their "neutrality" policy and join the West in fighting the Communist onslaught.

Notes

1. Dwight D. Eisenhower, *The White House Years*, vol. 1, *Mandate for Change* (Garden City, N.Y.: Doubleday, 1963), 459.
2. Chiou-Ling Yeh, "In the Traditions of China and in the Freedom of America: The Making of San Francisco's Chinese New Year Festivals," *American Quarterly* 56, no. 2 (2004): 395–420.
3. Yeh, "In the Traditions of China," 395–420.
4. Haiming Liu, "Transnational Historiography: Chinese American Studies Reconsidered," *Journal of the History of Ideas* 65, no. 1 (2004): 135–53.
5. Gary Klintworth, "Taiwan's International Identity: 400 Years in the Melting Pot," *Journal of East Asian Affairs (South Korea)* 10, no. 2 (1996): 373–94.
6. A. Hummel, ed. *Eminent Chinese of the Ch'ing Period* (Washington, D.C.: Government Printing Office, 1943), 108–10; I. Hsu, "Chinese Colonization of Taiwan" (Ph.D. diss., University of Chicago, 1975), 92.

7. H. Kessler, *K'ang-Hsi and the Consolidation of Ch'ing Rule, 1661–1684* (Chicago: University of Chicago Press, 1976), 90–97.

8. G. Tsai, "The Taiwan Straits Crises: Analysis of the China, Taiwan, United States Relationship, 1949–1983" (Ph.D. diss., Northern Arizona University, 1985), 10.

9. Bank of Taiwan, *Three States of Taiwan's Cession* (Taipei: Ko Tai San Chi, 1959), 29.

10. George H. Kerr, *Formosa: Licensed Revolution & the House Rule Movement, 1895–1945* (Honolulu: University Press of Hawaii, 1974), 68.

11. Jen-Kun Fu, *Taiwan and the Geopolitics of the Asian-American Dilemma* (Westport, Conn.: Praeger, 1992), 24.

12. Steve Phillips, "Confronting Colonization and National Identity: The Nationalists and Taiwan, 1941–45," *Journal of Colonialism and Colonial History* 2, no. 3 (2001).

13. Hungdah Chiu, ed. *China and the Question of Taiwan: Document and Analysis* (New York: Praeger, 1973), 203.

14. Qingdao Dashiji Shiliao, *A Chronicle of Important Historical Events in Qingdao,* vol. 1, trans. Shannon Wang (Qingdao: Qingdaoshi Danganguan, 1989), 46.

15. Chiu, *China and the Question of Taiwan*, 213.

16. United States Department of State, *American Foreign Policy, 1950–1955, Basic Documents*, vol. 2, (Washington, D.C.: Government Printing Office, 1957), 1667–68.

17. Steve Phillips, "Confronting Colonization."

18. Mingtong Chen, "Paixi Zhengzhi Yu Chen Yi Zhi Tai Lun" [A discussion of political factions and Chen Yi's rule of Taiwan], in *Taiwan Guangfu Chuqi Lishi,* ed. Jeh-Han Lai, trans. Shannon Wang (Nangang: Zhongyang Yanjiuyuan Zhongshan Renwen Shehui Kexue Yanjiusuo, 1993), 257–59.

19. M. Mancall, ed., *Formosa Today* (New York: Praeger, 1964), 4.

20. *ROC Yearbook* (Taipei: Government Information Office, 1988), 99.

21. Neil Jacoby, *U.S. Aid To Taiwan* (New York: Praeger, 1966), 46.

22. George Woei Tsai, "The Taiwan Straits Crises: Analysis of the China, Taiwan, United States Relationship, 1949–1983," (Ph.D. diss., Northern Arizona University, 1985), 20.

23. Hungdah Chiu, *China and the Taiwan Issue* (New York: Praeger, 1979), 81.

24. A. Amsden, "Taiwan's Economic History: A Case of Etatisme and a Challenge to Dependency Theory," *Modern China* 5 (1979): 341–80; I. Hsu, *The Rise of Modern China* (New York: Oxford University Press, 1983); T. Gold, *State and Society in the Taiwan Miracle* (Armonk, N.Y.: M. E. Sharpe, 1986).

25. *Important Documents Concerning the Question of Taiwan* (Beijing: Foreign Language Press, 1960), 121.

26. *Important Documents*, 123.

27. Waldo Heinrichs, "Eisenhower and the Sino-American Confrontation," in the *Great Powers in East Asia, 1953–1960*, ed. Warren I. Cohen and Akira Iriye (New York: Columbia University Press, 1990), 86–103.

28. Excerpts from part 2 of the Special Committee's Report on Southeast Asia, April 5, 1954; "Cablegram from Secretary Dulles to Ambassador Dillon in Paris, April 1954," *New York Times*, July 5, 1971.

29. Robert F. Randle, *Geneva 1954* (Princeton, N.J.: Princeton University Press, 1969), 322.

30. As early as 1952, the Chinese military began making plans for the capture of the offshore islets. De Wang, "The Battle to Liberate the Dachen Isles," in *San Jun Huizhan (The Land, Air and Naval Fighting in the East China Sea)*, ed. Fengzhi Nie (Nanjing: Jiefangjun Chubanshe, n.d.), 2.

31. George Woei Tsai, "The Taiwan Straits Crises," 207.

32. Neal Rosendorf, "John Foster Dulles's Nuclear Schizophrenia," pp. 62–86 in *Cold War Statesmen Confront the Bomb: Nuclear Diplomacy Since 1945*, ed. John Lewis Gaddis, Philip H. Gordon, Ernest R. May, and Jonathan Rosenberg (Oxford: Oxford University Press, 1999).

33. O. R. Holsti, "The Belief System and National Images: A Case Study," *Journal of Conflict Resolution* 6 (1962): 244–52.

34. Kwan Ha Yim, ed., *China and the U.S. 1955–63* (New York: Facts on File, 1973), 114.

35. P. Lyon, *Eisenhower: Portrait of the Hero* (Boston: Little, Brown, 1974), 646–47.

36. Alastair Iain Johnston, "Cultural Realism and Strategy in Maoist China," in *The Culture of National Security: Norms and Identity in World Politics*, ed. Peter J. Katzenstein (New York: Columbia University Press, 1996), 216–70.

37. Paul Godwin, "Change and Continuity in Chinese Military Doctrine: 1949–1999," in *Chinese Warfighting: The PLA Experience Since 1949*, ed. David M. Finkelstein, Michael A. McDevitt, and Mark A. Ryan (Armonk, N.Y.: M. E. Sharpe, 2001), 23–55.

38. Alexander L. George, David K. Hall, and William E. Simons, *The Limits of Coercive Diplomacy* (Boston: Little, Brown, 1971), 1–35; Allen S. Whiting, "China's Use of Force, 1950–96, and Taiwan," *International Security* 26, no. 2 (2001): 103–31.

39. Whiting, "China's Use of Force," 109.

40. United States Department of State, *Foreign Relations of the United States: Diplomatic Papers, 1952–1954*, vo. 14, *China and Japan*, (Washington, D.C.: Government Printing Office,) 556–57, 575–77.

41. United States Department of State, *Diplomtic Papers, 1952–1954*, vol. 14, *China and Japan*, 556–57, 575–77.

42. United States Department of State, *Diplomtic Papers, 1952–1954*, vol. 14, *China and Japan*, 583–88.

43. United States Department of State, *Diplomtic Papers, 1952–1954*, vol. 14, *China and Japan*, 577.

44. "President Says Atom Bomb Would Be Used Like Bullet," *New York Times*, March 17, 1955, 1.

45. Hou Tsung Chien, "U.S. Policy toward China, 1912 to Present Emphasizing Bureaucratic Level Analysis, Particularly Since 1936, of Both Nationalist and Communist Movements" (Ph.D. diss., Northern Arizona University, 1985), 222.

46. Thomas E. Stolper, *China, Taiwan, and the Offshore Islands* (Armonk, N.Y.: M. E. Sharpe, 1985), 34.

47. Whiting, "China's Use of Force," 109.

48. United States Department of State, *Foreign Relations of the United States: Diplomatic Papers, 1955–1957*, vol. 2, *China* (Washington, D.C.: Government Printing office), 83–85; Dwight D. Eisenhower, *The White House Years*, vol. 1, *Mandate For Change* (Garden City, N.Y.: Doubleday, 1963), 483.

49. Congressional Quarterly, *China and the U.S. Far East Policy, 1945–1967* (Washington, D.C.: Congressional Quarterly Service, 1967): 72.

50. Guoqiang Zheng, "Conflict, Contact, and Coexistence: Sino-American Relations, 1948–1960" (Ph.D. diss., University of Toledo, 1997), 391.

51. Tillman Durdin, "Zhou Asks for U.S. Talks on Formosa Crisis; Washington Sets Terms," *New York Times*, April 24, 1955, 1.

52. Kenneth T. Young, *Negotiating with the Chinese Communists: The United States Experience, 1953–1967* (New York: McGraw Hill, 1968), 38.

53. Cheng-Yi Lin, "China, the U.S., and the Security of Taiwan" (Ph.D. diss., University of Virginia, 1987), 12.

54. Lin, "China, the U.S," 13.

55. Lin, "China, the U.S," 13.

56. Lin, "China, the U.S," 14; Guoqiang Zheng, "Conflict, Contact, and Coexistence," 395.

57. M. Small, "Public Opinion," in *Encyclopedia of American Foreign Policy: Studies of the Principal Movements and Ideas*, ed. A. Deconde (New York: Scribner, 1978), 844–55.

58. Zheng, "Conflict, Contact, and Coexistence," 394.

59. Alexis Johnson, *The Right Hand of Power* (New York: Free Press, 1984), 246; Guoqiang Zheng, "Conflict, Contact, and Coexistence," 396.

60. Zheng, "Conflict, Contact, and Coexistence," 396.

61. Zheng, "Conflict, Contact, and Coexistence," 96.

62. Zheng, "Conflict, Contact, and Coexistence," 397.

63. Kwan Ha Yim, ed., *China and the U.S. 1955–63* (New York: Facts on File, 1973), 64.

64. Zheng, "Conflict, Contact, and Coexistence," 399.

65. Alice Langley Hsieh, "The Sino-Soviet Nuclear Dialogue, 1963."

66. Yawei Liu, "The United States according to Mao Zedong: Chinese-American Relations, 1893–1976" (Ph.D. diss., Emory University, 1996), 329.

67. Robert S. Norris, William M. Arkin, and William Burr, "Where They Were," *Bulletin of the Atomic Scientists* 55, no. 6 (November/December 1999): 26–35.

68. Tsai, "The Taiwan Straits Crises," 131.

69. Lin, "China, the U.S," 74.

70. Mao, "To Peng Dehuai," in "PLA Attacks and Amphibious Operations," from *Jianguo Yilai Mao Zedong Wengao* [The manuscripts of Mao Zedong since the founding of the nation], trans. Shannon Wang (Beijing: Central Documents Publishing Co., 1987–90), vol. 7: 326.

71. Tsai, "The Taiwan Straits Crises," 133.

72. Tsai, "The Taiwan Straits Crises," 133.

73. *ROC Yearbook 1955–56* (Taipei: Government Information Office, 1956), 4.

74. Lin, "China, the U.S," 76.

75. Dwight D. Eisenhower, *The White House Years*, vol. 2, *Waging Peace, 1956–1961* (New York: Doubleday, 1965), 293.
76. Dwight D. Eisenhower, *Waging Peace*, 293.
77. Alice Langley Hsieh, *Communist China's Strategy in the Nuclear Era* (Englewood Cliffs, N.J.: Prentice Hall, 1962), 124.
78. Lin, "China, the U.S," 76.
79. Allen S. Whiting, "Quemoy 1958: Mao's Miscalculations," *The China Quarterly* 62 (June 1975): 263–70.
80. Quoted in A. Doak. Barnett, "The 1958 Quemoy Crisis: The Sino-Soviet Dimension," *Problems of Communism* 25 (July/August 1976): 38.
81. Cited in Zhisui Li, *The Private Life of Chairman Mao*, trans. Tai Hung-Chao (New York: Random House, 1994), 261–62.
82. Lin, "China, the U.S., 77.
83. Kenneth T. Young, *Negotiating with the Chinese Communists: The U.S. Experience. 1953–67* (New York: McGraw-Hill, 1968), 179.
84. Lin, "China, the U.S," 80.
85. Whiting, "China's Use of Force," 103–31.
86. Whiting, "China's Use of Force," 110.
87. Wu Lengxi, "Inside Story of the Decision-Making During the Shelling of Jinmen," trans. Shannon Wang, *Zhuanji Wenxue* [Biographical Literature], no. 1 (1994), 11.
88. Lin, "China, the U.S," 79.
89. Lin, "China, the U.S," 279.
90. American Foreign Policy, *Current Documents, 1956–1967* (Washington, D.C.: Government Printing Office, 1958), 1207.
91. Congressional Quarterly, *China and the U.S. Far East Policy, 1945–1967* (Washington, D.C.: Congressional Quarterly Service, 1967): 101.
92. CQ, *China and the U.S. Far East Policy, 1945–1967*, 104.
93. CQ, *China and the U.S. Far East Policy, 1945–1967*, 105.
94. Lucian Pye, *China: An Introduction* (Boston: Little, Brown, 1978), 246.
95. Donald S. Zagoria, *The Sino-Soviet Conflict, 1956–1961* (Princeton, N.J.: Princeton University Press, 1962), 168.
96. Robert Sutter, *China Watch: Toward Sino-American Reconciliation* (Baltimore: Johns Hopkins University Press, 1978), 63.
97. Jasper Becker, *Hungry Ghosts: China's Secret Famine* (New York: Henry Holt, 1998).
98. Liu, "The United States according to Mao Zedong," 349.
99. Liu, "The United States according to Mao Zedong," 349.
100. Liu, "The United States according to Mao Zedong," 349.
101. M. S. Kapitsa, *On Different Latitudes—Diplomat's Notes* (Moscow: Kniga i biznes, 1996), 63.
102. Zedong Mao, "On the Ten Major Relationships," April 25, 1956, in *Selected Works of Mao Zedong* (Beijing: Foreign Languages Press, 1977), 5:288.
103. Liu, "The United States according to Mao Zedong," 354.
104. Guoqiang Zheng, "Conflict, Contact, and Coexistence," 489.

105. Philip Wheelwright, *The Burning Fountain: A Study in the Language of Symbolism* (Bloomington: Indiana University Press, 1968), 7.

106. Roberta Cohen, "People's Republic of China: The Human Rights Exception," *Human Rights Quarterly* 9, no. 4 (1987): 447–549; see also Susan Shirk, "1977–1978. Human Rights: What about China?" *Foreign Policy*, no. 29 (Winter 1987): 109–27.

107. Adam J. Zweiback, "The 21 'Turncoat GIs': Nonrepatriations and the Political Culture of the Korean War," *Historian* 60, no. 2, (1998): 345–62.

CHAPTER 4

Interlude and Stasis, 1960–1967

> "[W]hy is the Soviet Union interested in talking to the U.S. in a number of fields? While they of course jump through the ceiling if you talk about linkage, they link everything. What are the Soviet Union's problems? They look at China—they know they have nothing to fear now from China, but they also know what they themselves have done economically in fifty years in a relatively backward country. They respect the Chinese people. They know that a billion Chinese in twenty-five years could be an enormous threat in the future."
>
> President Richard M. Nixon, June 16, 1972[1]

1. Introduction: The Long Handshake

Dwight David Eisenhower ended his presidency as more of a China hawk than when he entered office. The opportunity in 1956 for some sort of rapprochement or at least Sino-American–style mutual recognition was long dead.[2] In fact, Ike told his successor, John F. Kennedy, that he would not interfere or speak against the new commander in chief on any issue except if the White House endorsed the PRC's admission to the United Nations.[3] That said, Ike was aware of the mixed feelings and misgivings in the public about a permanent break with mainland China because he held them himself. According to Stephen Ambrose's biography of the president, after the end of the Korean War Eisenhower felt that "keeping China out of the U.N., and refusing to recognize the existence of the Communist government, made no sense. But the art of leadership includes the art of the possible, and Eisenhower believed that the American public was not ready to think about a new relationship."[4] Another Eisenhower biographer, Robert J. Donovan, emphasizes Eisenhower's suspicions of China and his failure to see any great advantage to the United States in the PRC's recognition.[5] Nor was the president pressured by any "recognition bloc" on Capitol Hill, in the press, or among the general population. In sum, the status quo seemed to be the safest position.

Who spoke for the other side? Edgar Snow, who penned the best book about Mao and Communists of his generation, *Red Star over China*, was still active—but not visiting China. There were other voices, and even in *Time* they made themselves heard—but not much. The three confrontations with China—the Korean War and the two Taiwan Straits crises—along with an anti-Communist fervor at home had left few of the more sympathetic "old China hands" in government service or willing to speak out openly.[6] Of these, supporters of the PRC were very few: the 10 percent in public opinion polls who advocated recognizing the mainland and dumping Chiang were not Maoist partisans. In fact, even after all the conflicts with the PRC, according to some polls as many as a quarter of Americans did not even know that China was a *Communist* country. And even fewer Americans would have had any inclination to admire Maoism, whereas not a few romanticized Hitler, Mussolini, and Stalin in a previous era. Only one publisher in America distributed PRC literature in translation, and he was harassed by the FBI.[7] Indeed, Mao had no real constituency in America in 1960, but some Americans simply felt, as did Ike at times, that rapprochement made sense, although not at a limitless price. There was one flicker of daylight in the late 1950s, when a number of prominent senators and congressional representatives expressed the need for a new look at America's relations with the PRC,[8] but they were just making a statement, not forming a voting bloc, and their musings faded away. Indeed, the late 1950s and early to middle '60s were a fallow period for anyone taking the initiative in China policy in American politics.

The two presidents following Eisenhower—Kennedy and Johnson—also had misgivings about the nonrecognition of the PRC, its exclusion from the world body of nations, the microscopic American trade with the mainland, and the fiction that tiny Formosa housed the authentic government of China. But for most of the 1960s little changed in actual policy or even much in the content of the invective flying across the Pacific. In fact, it could be said that Truman, Eisenhower, and the two '60s Democrats all felt that something was wrong with the U.S. relationship with China but found that circumstances—some of their own creation, some forced on them—allowed changes only for the worse. The world was so used to rigidity in Sino-American relations that it would be shocked that Eisenhower's vice president, an old Cold Warrior and longtime friend of Chiang and his cause, would be the first American president to set foot on the mainland and grasp the hand of Mao Zedong in the Great Hall of the

People. It was a dramatic leap across a divide symbolically and historically greater than the distance between Washington and Beijing. As William F. Buckley, founder of the conservative *National Review* (and who accompanied the president to China), noted with tongue in cheek, "[Nixon] would have had a hell of a time explaining it to the Committee on Un-American Activities."[9]

Nixon's move seemed startling in contrast to the "static and frozen" previous decades of Sino-American relations, where hundreds of talks, thousands of speeches, one war, and several crises resulted in only one signed agreement (on prisoner exchanges in 1955).[10] But such a projection of punctuated equilibrium, that is, stasis broken by dramatic change accomplished by Occidentals, masks a complicated timeline of crossed agendas in the 1960s. Many events and ideologies led toward the road to rapprochement. It was not a series of fixed dominoes that would result in the photo icon of "Nixon in China" but rather contingencies that might have veered in varied directions. And, of course, we must be careful about ascription of the active tense to the West: Nixon did not "win" China, just as America had not once "lost" it. Nixon went to China, but China invited Nixon to come. Mao could have hung up the "No Visitors" sign for another presidency or another decade. And, notably, as is often the case with great events, China experts in the West, like academics, were poor prophets. Publications by sympathetic Chinese scholars as late as 1971 contained no hint that rapprochement was imminent.[11] Nor, as described below, were journalists any more sensitive to the great changes in the wind.

On the other hand, many forces favored some sort of breakthrough, and now, through the crystal ball of hindsight, we can easily identify them. China had passed through great traumas such as famine, a self-destructive Cultural Revolution, purges of top leaders, conflicts with neighbors, and missteps in diplomacy. A new enemy loomed as well. The Soviets had hoped that when Khrushchev was purged and a new leader came to power in Russia, relations with China might take a turn for the better; it was not to be. Despairing of Mao's polices, the USSR hoped for a change in leadership for their southeastern neighbor. Former Soviet deputy foreign minister for the Far East M. S. Kapitsa recalls that in 1965 his country's defense minister told Zhou Enlai, "We got rid of our fool, [now] you get rid of yours."[12]

In 1969, the Soviet Union and China engaged in border clashes. The Soviets even made noises (and inquiries) about attacking China and wiping out its nuclear capability. America was still viewed by China as an imperialist but perhaps not a threat worth walling out. Character counted, as well. Mao trusted tough opponents: Nixon's Cold War credentials are often cited as allowing him to escape political fallout domestically from a China rapprochement, but it is forgotten that the same résumé differentiated Nixon, in Mao's estimation, from the "mixed signals" presidents of the past, such as Truman and Johnson. Mao's physician recalled that after the chairman met Nixon, he expressed appreciation that the American president had explained the benefits of improved Sino-American relations on a practical level and had not resorted to appeals to "high moral principles."[13]

In the United States, by 1969, the year Nixon took office, a growing host of louder voices in Congress, in the academy, and elsewhere were urging reconciliation between the great nations.[14] The seeds of rapprochement were "in the air."[15] The China bloc had atrophied; the Committee of One Million, its leading lobby group, was a shell. Many of the famous Chiang-supporting senators and congressmen had retired or passed away—although a strong pro-ROC element survived on the Republican right. In the spring of 1964, Henry Luce stepped down as editor in chief of his empire's publications. His successor—not given the same title—was Hedley Donovan, who, while loyal to Luce, chafed at many "Luce-isms," such as the rhetorical barbing of enemies and the maintenance of the party line.[16] America was mired in a war in Indochina, and many assumed that the road to peace with Hanoi led through Beijing. The United States desired an accommodation on a number of issues with the USSR (what would become détente), but felt some sort of pressure (what would become the "China Card") needed to be exerted to motivate the Soviets to accommodate.

Finally, as we shall see, one major factor in policies toward China in the 1960s and early 1970s was one of the oldest visions of China, one remarked upon by Marco Polo: the Middle Kingdom's "mass." For the Russians at that time, China's vast population was a yellow-periled threat. Khrushchev had described China as "a mass of human flesh and nothing else"; his successors feared nukes as well as the hordes when China tested its first atomic device in 1964, accomplished its first combined launch and detonation of a nuclear-armed missile in 1966, set off its first hydrogen bomb in 1967, and fired its first intercontinental ballistic missile in 1971.[17] For Chinese leaders, the "mass" was a political threat (a possible revolt at failed domestic policies) but also an opportunity (for mobilization against internal and external enemies). For American policymakers, China's hundreds of millions could still be perilous. President Kennedy once commented, "I would regard that combination of weak countries around it, seven hundred million people, a Stalinist internal regime, and nuclear powers, and a government determined on war as a means of bringing about its ultimate success, as potentially a more dangerous situation than any we faced since the end of the Second [World] War."[18]

More vital, as we shall see below, fear of those masses, borne by memory of Korea, would guide much of America's policies in Asia, and especially in Vietnam. In an interview conducted in 1969, Walt W. Rostow, chairman of the State Department Policy Planning Council (December 1961–April 1966) and special assistant to the president for national security affairs (April 1966–January 1969), commented that once Kennedy had decided to take a stand against Communism in Southeast Asia, "He'd rather fight in Vietnam than in Laos. . . . In Laos you're on the Chinese border. In Vietnam you've got the cushion of North Vietnam between you and China. Therefore there was less chance, if you defended Southeast Asia basically in Vietnam, of getting into a war with China."[19] Over and over, aides to President Johnson would also cite "not wanting to get into a war with China" as the leading influence on Vietnam policy. Some would see the avoidance of that fate as a major accomplishment; others, especially in the military, would grumble that Sinophobia drove America into fighting an interminable, limited, and unwinnable war.

But the symbolism of the immensity of China's population would finally serve those seeking reconciliation, even on the most basic level, with China. President Nixon would cite the Chinese mass on July 15, 1971, in a live television address to the nation that "shocked the world" by revealing his planned travel to Beijing: "I have requested this television time tonight to announce a major development in our efforts to build a lasting peace in the world. As I have pointed out on a number of occasions over the past three years, there can be no stable and enduring peace without the participation of the People's Republic of China and its 750 million people. That is why I have undertaken initiatives in several areas to open the door for more normal relations between our two countries." When Nixon would eventually tour China, a new generation of Americans would become fascinated by the massiveness and the masses of the exotic land revealed (but only partly) via images. He would create a photojournalistic icon by shaking Mao's hand—the very same hand referred to by *Time* magazine, sixteen years previously, as follows: "Because it is hard for the mind to visualize so vast a slaughter in human terms, the Communists have been able to reap an advantage from the very size of their funeral pyre: many Westerners, finding the monstrous incredible, cannot see the blood on the hand of pretended friendship proffered by Chinese Communist leader Mao Tsetung." *Time* in 1972, however, would approve of the handshake.

2. Kennedy: Wars of Words

Because John Fitzgerald Kennedy died in office, we are not left with the extensive remembrances typical of other presidents, in memoirs, postpresidential interviews, and documents. JFK historiography tends to be divided on the question of the if-he-had-lived trajectory on foreign and domestic policy. Would Kennedy have become a Nixon, working toward rapprochement with the PRC, perhaps flying to Beijing? Would Kennedy have withdrawn from Indochina and avoided the tragedy of American intervention enacted by his successor? On the latter question, President Johnson himself appealed to the Kennedy legacy and precedents to justify his escalation in Vietnam.[20] After his tenure, LBJ reminisced, "Everything that I had ever learned in the history books taught me that martyrs have to die for causes. John Kennedy had died. But his 'cause' was not really clear. That was my job. I had to take the dead man's program and turn it into a martyr's cause. That way Kennedy would live forever and so would I."[21] But was the "cause" a crusade in the jungles of Indochina? Certainly JFK, while running for president, sounded like the staunchest Cold Warrior, with only a few expressions of moderation. His main thumping point was, after all, a "missile gap" (which turned out to be spurious). In office, Kennedy made many statements that today would mark him as a true believer of the right, and his decision making on most foreign affairs issues did not speak of someone seeking to withdraw from commitments, such as that to South Vietnam.[22]

On the other hand, many of Kennedy's aides, especially those who were more liberal themselves—such as Arthur M. Schlesinger, the president's special assistant, and Theodore Sorensen, special counsel to the president (and ghostwriter of Kennedy's Pulitzer Prize-winning book *Profiles in Courage*)—painted a president who, in a second term, would make bold changes, especially in American foreign policy. Another aide, James Thomson, would later claim that the Nixon China visit "shock" was made possible by JFK's actions and policies.[23] Robert Kennedy would contend in the 1968 presidential election that Johnson had abandoned JFK's prudence on Indochina.[24] The Kennedy second-term counterfactual has even become part of popular culture: In Oliver Stone's film *JFK*, the martyred leader's plan to pull out of Vietnam is the impetus for the military-industrial complex (and by heavy implication Johnson) to assassinate him. But on China, the polarities are less severe because the issue seems more abstract. Kennedy wavered in attitudes toward the PRC and thus, as outlined below, quotes, actions, and observations can be plucked to support different futures. One notation made by Schlesinger, however, is important, although it does not resolve any of the basic issues. Kennedy was one of the most take-control presidents in foreign policy in America's history; he had no elephantine Dulles to push him, and the president was "increasingly the day-to-day director of American foreign policy."[25]

In his earliest political expression about China, Kennedy comes off as a tough anti-Communist with a hint of pragmatic optimism.[26] His statements in the late 1940s about the "loss of China" read as if they could have been made by an anti-Truman Republican. As a congressman in 1948 he stated, "Our relation with China since the end of the Second World War has been a tragic one . . . and it is of the utmost importance that we search out and spotlight those who must bear the responsibility for our present predicament."[27] By the beginning of the next year he had found the guilty party: "The responsibility for the failure of our foreign policy in the Far East rests squarely with the White House and the Department of State."[28] According to an unconfirmed but credible story, Kennedy even donated $1,000 to the campaign of a young fellow navy officer named Richard Nixon running against a very liberal Democratic opponent.[29] In the '50s, he described U.S.-PRC economic relations as "trade in blood."[30] In an article written in 1957, he expressed agreement with Eisenhower's nonrecognition policy but added, "We must be very careful not to straightjacket our policy as a result of ignorance and fail to detect a change in the objective situation when it comes."[31] But a clearer expression of his thoughts may have arisen in his famous debates with Richard Nixon. The central issue of China at that time was not trade, recognition, or even the seating of China in the United Nations, but Jinmen and Mazu. In the second Kennedy-Nixon debate, the senator said, "I think it's unwise to take the chance of being dragged into a war which may lead to a world war over two islands which are not strategically defensible."[32]

Such statements convinced some in the Eastern bloc that Kennedy would be an improvement over Eisenhower's vice president as an adversary. In one of his memoirs, Khrushchev recalled that he had a generally favorable impression of Kennedy, mainly because in one of the presidential debates he said that he would have, unlike Eisenhower, apologized for the U2 incident (in which an American spy plane was shot down over the USSR). Overall, the former Russian premier noted, "We had little knowledge of John Kennedy," other than that he was "a young man, very promising and very rich—a millionaire . . . distinguished by his intelligence, his education, and his political skill."[33] Khrushchev, in their summit of June 1961 at Vienna, joked with the American president that he had "voted" for him by not releasing captured U2 pilot Gary Powers until the day after the election.[34]

Upon his victory, Kennedy appointed a number of men such as Chester Bowles (special representative and adviser on African, Asian, and Latin American affairs) and Adlai Stevenson (ambassador to the United Nations) who advocated flexibility in America's China policy. Bowles, for example, recommended a clear message to Chiang Kai-shek to drop his pretense that he was the real president of China.[35] Stevenson's position toward the PRC was even more conciliatory. While not urging immediate recognition, he felt that the United States should accept China's application for UN membership and prod the Nationalists to evacuate the offshore islets.[36] Neither man, however, championed China as a priority for the first year's agenda of the new presidency; there is no evidence that anyone in the JFK administration did. (In comparison, Nixon did not fly to Beijing until the *third* year of his tenure.) In any case, other events blacked out the foreign policy radar and seemed more pressing. In late December, foreign policy crises erupted all over the world, including in Indochina, Cuba, the Congo, and Berlin. The China problem would have to be put off.[37]

The Chinese themselves helped sabotage any possibility of an early Kennedy handshake. The summer of 1960 was a time of many domestic upheavals, bad economic news, and foreign policy reversals for the PRC. The Sino-Soviet split continued to widen, and the two Communist nations began to vie for supremacy in the Communist bloc. Widespread crop failure, forced communization, and the Great Leap Forward's collapse had caused a famine, with the dead numbering in the tens of millions. The period of 1959–1961 became known as the "three bitter years" (*San k'u-nien*).[38] That the disaster was not natural but a politically driven failure of policy is now undisputed.[39] There were some attempts, however, to signal openness to the Americans. For example, at an international conference on August 1, 1960, Zhou "advocated that the countries in Asia and around the Pacific, including the United States, conclude a peace pact of mutual non-aggression and make this entire region a nuclear weapon-free zone."[40] Likewise, on October 1, during the celebration of the eleventh anniversary of the founding of the People's Republic, Zhou and Foreign Minister Chen Yi both spoke in favor of improved international relations.

Autumn, however, brought yet another rejection of China's application for UN status. Chinese propagandists singled out Nixon as an archenemy but expressed no great support for Kennedy.[41] Chagrined that its (modest) outreaches were not accepted, and possibly perplexed at the American electoral system in which a new president did not actually take power on election day, China launched a series of vitriolic attacks on Kennedy personally and the United States in general.[42] Perhaps as an attempt to reinforce to the Chinese people that the regime that had failed them on the level of basic economics was fending off an implacable external enemy, on the day before Kennedy's inauguration a major Chinese newspaper assured its readers that the "change in the U.S. administration does not signify any change in its policy."[43] Kennedy, who had perhaps the least foreign policy experience of any recent president since Truman, paid special, some would say inordinate, attention to the insults heaped on him by Beijing. Some of his aides tried to explain away the rhetorical missiles as impotent raging, but Kennedy took them as revelations of the heart.[44]

The Warsaw talks also offered little hope for any thaw from Beijing. Once again, the United States wished to deal with its most pressing (and most resolvable in its own eyes) issue of Americans held in China. The Chinese telescope was on the status of Taiwan and the presence of American forces on the island. The situation was effectively summed up, shortly after President Kennedy took office, in a meeting between John M. Steeves, deputy assistant secretary for Far Eastern affairs, and the ambassador of the Republic of China. The Americans reassured the Taiwanese that they "could not see any prospect that this question [of the recognition of the PRC] would arise in any form which would make such a development possible. There was no indication from Beijing that the Chinese

Communists were interested in bringing this about, and we ourselves had no intention of taking any initiative. . . . Beijing says that the United States has to abandon his [Chiang's] government. We, of course, will not do that."[45] And so it came to pass. Indeed, the reports of the U.S. representatives at the Warsaw ambassadorial talks read like a psychologist's minutes of a meeting with married couple on the verge of divorce. Reports of meetings are full of petty recriminations.

Kennedy had a long list of other reasons to avoid a China initiative of any seriousness or boldness. Another Kennedy historian quoted the president as believing that "any American initiative toward negotiation, diplomatic recognition, or U.N. admission would be regarded by the Chinese as rewarding their aggressiveness."[46] The Far Eastern section of the State Department consisted of hardliners and silent moderates who, recalling the purging of the China hands in the late 1940s and '50s, had neither inclination nor incentive to advocate any overturning of American policy.[47]

During the period of 1961–62, Schlesinger projected that Kennedy's main worry about a rapprochement with China was domestic politics.[48] In a March 3, 1961, meeting with the prime minister of New Zealand, Kennedy candidly admitted that "China questions must be dealt with in such a way as not to cause deep divisions within our nation, and he [Kennedy] mentioned former President Eisenhower's strong views on the China question and the risks that a national debate on the subject might entail, including the risk of mounting isolationist sentiment."[49] On paper, Congress was still radically against any rapprochement with China. A 1961 resolution opposing "the admission of Communist China to the United Nations" was adopted by both houses with no recorded "nays."[50] As late as 1964, 345 congressmen were on record supporting the Committee of One Million.[51] And then there was Henry Luce; Kennedy's personal relationship with the publisher was occasionally on the level of screaming and table pounding, but the president was, as noted, "obsessed with *Time*."[52] The politically savvy JFK must have known that challenging the China consensus would set off a crusade against him in the Lucepress.

Finally, Kennedy had won election by a narrow margin of one hundred thousand votes; suspect voting practices in Chicago call into question whether he actually won at all. Kennedy was treading carefully through policies both foreign and domestic in an attempt to avoid alienating his not-overwhelmingly-supportive citizenry. Opinion polls on the China question were mixed. On the one hand, the American people, when asked about specific points on China policy, seemed downright moderate. Between February 10 and 15, 1961, Gallup conducted a national survey that asked, "There is now a severe food famine in Communist China. Do you think the United States should or should not offer to send some of its surplus foods to this country?" 52 percent of the respondents said that the United States should send the food, while 37 percent opposed the hypothetical offer. On whether the United States should "work out a business arrangement to buy and sell goods" with Communist China, 47 percent approved; 35 percent opposed. The sample was also asked, "Looking ahead to 1970, which country do you think will be the greater threat to world peace—Russia or Communist China?" 49 percent of the respondents listed Russia; 32 percent answered China. Of those who answered "China," when asked why that country "will be the greater threat," the respondents answered with the following itemized breakdown: "huge population, needs more land, food" (33 percent); "warlike policy, disregard for human life" (29 percent); "developing industry, will have nuclear weapons" (14 percent); "backward, ignorant people, can be led into war" (11 percent); "dislikes U.S., not recognized by U.S." (5 percent). On the other hand, 59 percent of the same group of respondents answered that the United States "should go along" if the United Nations ended up admitting Communist China as a member. The opposition to UN membership for the PRC, however, stood at about its usual level of 64 percent, while 53 percent of the sample thought the United States should "take steps to improve our relations with Communist China."

In 1962, 48 percent of respondents to a Gallup survey taken between May 31 and June 5 supported sending food to the PRC while 43 percent opposed. In 1963, a survey conducted between March 8 and 13 by the Gallup organization asked its respondents to look ahead to 1970 and assess "which country do you think will be the greater threat to world peace—Russia or Communist China?" The same number as resulted in the 1961 survey—32 percent—listed Communist China. In sum, the China issue seemed low on a long list of possible areas that were likely to create a political firestorm.[53] Kennedy had no mandate for overturning a bedrock of American policy and few partisans to help him do the lifting, even if he had wanted to.

A number of Kennedy aides, however, point to certain statements made by the president that presaged that he was hardly an implacable Dulles. Theodore Sorensen speculated that a second term might have produced a different China policy, for Kennedy was not dedicated to perpetual "rigidity" on the subject.[54] Kennedy thought there might come about internal changes in China that would reward future American steps toward reconciliation: "He hoped that the passage of time, an evolution among Red China's leaders, their isolation from the rest of the world, their mounting international problems, and their inability to gain through aggression would be persuasive as well."[55] The president further commented to Adlai Stevenson, "It really doesn't make any sense—the idea that Taiwan represents China."[56] Another administration official, Roger Hilsman, who was assistant secretary for Far Eastern affairs, projected an even more aggressive diplomacy in a second term, arguing that "President Kennedy would have liked to see progress in the same direction [détente with the Soviet Union] with Communist China."[57] Whether this was truth or retrospective wishful thinking we shall never know.

In office, Kennedy did deal with one major China incident—a third straits crisis—deftly and effectively. In this case, as before, Chiang Kai-shek was the wild card in Sino-American relations. In fact, June 1962 was a period of increased tension. Chinese refugees fleeing to border countries spoke of hunger, death, and chaos at home. Taiwan, naturally, perceived the problems in the PRC as an opportunity, and ratcheted up its "liberation" bombast. Chiang believed that the Communist regime was rotten and would collapse if given a kick from his own forces.[58] On its part, the PRC leadership once again detected American manipulation behind Chiang's saber rattling and decided that another demonstration of preventive aggression in the Taiwan Straits was necessary. A *People's Daily* editorial on May 19 announced: "We must serve a fresh warning to the Kennedy administration that it shall be held fully responsible for all grave consequences arising from its policy of playing with fire."[59] Seven People's Liberation Army divisions moved to the coast by the end of June, and China warned the United States through diplomatic channels that it would repel any invasion from Taiwan. U.S. intelligence estimates concluded that the new coastal deployment was not an invasion army but rather was designed to defend against landings on the coast. Kennedy responded quickly by reassuring Ambassador Wang Bingnan that the United States had no intention of launching any attack or allowing the Nationalists to do so. Privately, the Chinese accepted Kennedy's word, but publicly they continued to charge that American imperialism was aiming toward their country.

Yet another sign of caution on Kennedy's part came in the fall of 1962 during a border war between China and India. Since the late 1950s, Chinese troops, in support of their control of Tibet, had occupied a small portion of India's northwest territories that the PRC claimed were within China's borders. Talks between the two countries broke down, and on October 20 the PLA attacked at several points along the 2,500-mile border. After a pause, waiting for India to back down (which it did not), the PLA resumed its attacks, absorbing all the area it claimed to be under China's control. Nehru of India appealed for outside help but was cut short when China declared a unilateral cease-fire.[60] Kennedy's administration, as well, held back from real involvement in the crisis.

On the other hand, there were signs that a second-term Kennedy would be a continuing hardliner.[61] For example, against his debate rhetoric that offshore islands like Jinmen were not worth fighting for, Kennedy put no pressure on Chiang to evacuate them. He accepted the verdict of his special military representative, Maxwell D. Taylor, who reported in the findings of a Far East mission in September 1962, that "All American officials contacted in Taiwan agreed that the U.S. should recognize, however reluctantly, that the [Chinese Nationalists] will continue to hold and defend the Offshore Islands. The Islands are valuable as outposts for the main defense of Taiwan and the Penghus, as the site of early warning radar and as a symbol of the determination of the [Chinese Nationalists] some day to return to the mainland."[62] Other moves and statements suggested a tough line against China. JFK authorized a 66 percent increase in deployment of nuclear weapons to the Pacific theater.[63] He seriously considered joining the Soviet Union in a destruction of all Chinese nuclear facilities, launching sites, and research bases.[64] The CIA supported the Tibetan resistance. Also, the stonewalling of Beijing promised that even if Kennedy had been inclined to try to rally support for a new China policy, there were no signs from China that they were interested in the same, and the domestic hornet's nest was not worth disturbing.[65] Even Schlesinger admits that by 1963, Kennedy was tired of the caustic rhetoric emanating from Beijing and, we presume, from the transcriptions of the ambassadorial talks, and had concluded "that China presented the long-term danger to the peace."[66]

Yet, the skillful avoidance of war—and of escalation—suggests a Kennedy marked by prudence and eschewing adventurism. He had turned down the "surgical strike" option against Chinese nukes and still held open the policy of "containment without isolation" advocated by Roger Hilsman and his allies.[67] Also, as historian Lawrence Freedman has pointed out, in Indochina, the Berlin crises, and the Cuban missile crisis, "catastrophe was avoided"; this paints a president who "would have looked hard" to find a way to avoid the abyss of Vietnam—and not gotten sucked in as did his successor.[68] Then there was the fact that Kennedy, unlike many more experienced statesmen, seemed to actually learn from mistakes. The failed Bay of Pigs invasion made him even more dubious about Chiang's prospects for a return to the mainland. Meeting with Chiang Ching-Kuo, the Generalissimo's son and political heir on September 11, 1963, in Washington, Kennedy politely put off or rejected Nationalist proposals to step up raids on the mainland.

In short, Kennedy was unwilling to tell the Nationalists to drop their dream of reunification, but he knew it was dead and that he must keep it in its grave. It may be a stretch of credulity to praise JFK, a reckless man in his personal life, as a cautious one on public policy. Such a Kennedy may not have become the "liberal" his partisans of 1963 and today hope that he really was and would have been in a second term.[69] Yet, that habit of practical reason, admitting errors, thinking out of the box, and of not wanting to ever be caught in any more "losing operations" was exactly the JFK legacy Robert Kennedy appealed to when he ran for president in 1968 *against* the war in Vietnam.[70]

3. Johnson: "Don't Start a War with China!"

Lyndon Baines Johnson entered the office of the presidency, after Kennedy's assassination in 1963, with the sympathy of the nation and high approval ratings.

Then, using a highly suspect "attack" on American ships by North Vietnamese forces in the Gulf of Tonkin (between Vietnam and China) on August 7, 1964, he extracted a virtually unanimous resolution from the House and Senate to allow him to "take all necessary measures to repel any armed attack against forces of the United States and to prevent further aggression." In the election that year, he won in a landslide. His margin was due, in part, to the perception that he would keep America out of war in Vietnam, unlike the saber-rattling Barry Goldwater.[71] Indeed, as the "peace" candidate he had affirmed, "[We] don't want our American boys to do the fighting for Asian boys. We don't want to get involved in a nation [China] with seven hundred million people and get tied down in a land war."[72] But within a year, American troop levels in Indochina reached two hundred thousand in contrast to Kennedy's sixteen thousand advisers. Until the withdrawal from Vietnam under Nixon, America would lose nearly sixty thousand men in the undeclared war and nearly be torn apart over the issue at home. Each step of escalation would be portrayed by Johnson as a sacred continuation of the polices of his predecessors, most notably John F. Kennedy. The contradiction that Johnson was plunging America into a land war in Asia took years for even a majority of Americans, let alone political leaders, to realize because of the administration's step-by-step path into the abyss. My own father commented to me years later, "I voted for Johnson thinking he was the peace candidate, then found out he was the piecemeal president." In a famous column of 1965, humorist Art Buchwald wrote, "I had a nightmare that Goldwater was president and we were escalating the war in Vietnam."

Yet, in the case of China, Johnson was much more eager than Kennedy to break the impasse. Lawrence F. O'Brien, special assistant to Presidents Kennedy and Johnson, explained in an interview many years later that Johnson wanted unanimity in foreign policy, especially in light of the "loss of China" debate.[73] In psychological terms, LBJ was a man of immense insecurity, with a fear of failure, a talent for endless second-guessing of any decision, and a thirst to be loved and to win universal approval.[74] (He was incensed, for example, that the Tonkin Resolution received two "nay" votes in the Senate.) These qualities drove him to micromanage a war effort that—in the football analogies he loved so much—was nothing but a "prevent defense." Worse, he deceived the public into believing that victory was a goal and, as historian and Vietnam veteran F. Charles Parker put it, "left an army in combat whose only goal was to avoid defeat."[75] But the positive side of his internal demand for approval was that he felt that bridging the chasm to China was a personal imperative. And unlike in Eisenhower's or Kennedy's time, now the advocates for a new China policy were louder and more powerful. By 1965, the periodic vote on the admission of China to the United Nations had come down to the American veto, with even Britain and France supporting the PRC's admission.[76] In 1966, the Senate Foreign Relations

Committee held hearings on China policy, and a series of speakers, including Republican Nelson Rockefeller, Nixon's primary opponent in 1968, and Nixon's Democratic opponent Senator Hubert Humphrey, all spoke for movement toward rapprochement with China.[77] In contrast, by this time the China Lobby (supporting the Nationalists in Taiwan) was a ghost.

Pubic opinion, too, was mixed, although not "pro-PRC" by any means. Dean Rusk, who also served as Johnson's secretary of state, noted later, "I think that if it were possible to find a reasonable basis on which to improve relations with mainland China that the American people would be glad to see it happen."[78] In 1964, China's news profile was heightened by the increasing attention to the conflict in Indochina and China's development of atomic weapons. Unsurprisingly, in an April 24–29 Gallup survey, when the respondents were asked, "Which country do you think will be the greater threat to world peace—Russia or Communist China?" 56 percent answered Communist China and only 27 percent named Russia. A January 30–February 4 survey found 71 percent of respondents opposing the admission of Communist China to the UN and only 42 percent supporting the notion that the United States "should go along" if China were admitted despite U.S. opposition.

A Gallup survey of September 16–21, 1965, asked Americans, "What do you think is the most important problem facing this country today?" Vietnam was listed by 19 percent of the sample (the second-highest total for any issue), but "Red China" was cited by only 2 percent of the sample. A February 19–24 survey asked the perennial question about UN membership. The opposition number now was 64 percent. When those who opposed admission were asked their reasons why, these included the following: "Red China not trustworthy, not live up to rules" (13 percent); "no Communist country should be in the U.N." (11 percent); "admission would strengthen the Communist bloc" (10 percent); "admission would be liability to U.N." (9 percent); "they're too warlike, do not want to negotiate" (3 percent). Another 21 percent of the admission opponents listed other reasons or gave no particular reason. At the same time, the "should go along" faction among respondents asked what the United States should do if China were admitted to the UN stood at 49 percent.

In 1966, 67 percent of the respondents to a Gallup poll conducted December 11–16 opposed China's admission as a member of the United Nations. 55 percent of respondents surveyed March 24–29 believed that Communist China "should not be admitted as a member of the U.N." 49 percent of the respondents believed, however, that the United States should "go along" if the UN decided to admit the PRC. The same survey asked people to rate certain words on whether they described peoples from other countries best. When asked about "the Communist Chinese people," the respondents listed many characteristics: "hardworking" (37 percent); "ignorant" (24 percent); "warlike" (23 percent);

"sly" (20 percent); "treacherous" (19 percent); "intelligent" (14 percent); "religious" (14 percent); "cruel" (13 percent); "artistic" (13 percent); "ordinary" (12 percent); "radical" (9 percent); "practical" (8 percent). Gallup, when querying a *Who's Who in America* sample in a September 8–13 survey, asked, "Do you think Communist China should or should not be admitted as a member of the United Nations?" Fully 64 percent of the group supported the PRC's admission. Interestingly, when the general public was asked about China's admission to the UN and the corollary "if it would improve United States-Communist relations" was added, the group in support of Chinese admission increased from 25 to 55 percent.

In 1967, in a Gallup survey dated February 16–21, Americans "looking ahead" were asked whether Russia or Communist China "will be the greater threat to world peace." 71 percent named China; 21 percent cited Russia. Intriguingly, Sino-Soviet tensions were now notable to the American public; 48 percent of the sample felt that in a war between the United States and China, "Russia would be more likely to be on our side." A 1968 survey by Gallup taken January 4–9 asked about American attitudes toward certain foreign nations. Of all those listed, the least positive favorable rating, at 5 percent, was that of Communist China; in contrast, North Vietnam received a 7 percent rating and Russia a 19 percent rating. In sum, regarding public opinion from 1964–1968, while Johnson apparently had an opening to do something different with China, especially in the early years of his high popularity, there is no evidence of any strong pro-PRC sentiment in the public.

The most pressing reason to reach out to China, however, was peace. Unsure of how to resolve the Vietnam quagmire, Johnson began to feel that the road to compromise from Hanoi might lead through Beijing, and so a reconsideration of China policy was an obvious opportunity. China also loomed large in LBJ's mind because, if his operations in Vietnam had one dictum, it was "Don't start a War with China!" Johnson alone cannot be credited (or blamed) for this policy obsession: the evidence is overwhelming that Sinophobia permeated the administration. A May 5, 1965, National Intelligence Estimate on "Communist China's Foreign Policy," submitted by the director of the Central Intelligence Agency but prepared with the cooperation of other intelligence agencies, concluded: "To date, the Chinese leaders have not made risky countermoves to the limited U.S. attacks in North Vietnam, and they almost certainly seek to avoid a wider war. Nevertheless, they have been making preparations for at least limited engagement, and we believe that they would be prepared to risk a major military conflict with the U.S. should they feel China's vital security interests threatened by U.S. actions."[79] A memorandum from the under secretary of state for political affairs to Secretary Rusk on July 15, 1965, argued that "the Communist Chinese might become actively involved in the war if the Communist regime in

North Viet Nam were threatened, if Chinese borders were placed in jeopardy, or if large Chinese Nationalist forces were brought to the Southeast Asia mainland."[80] In an interview in 1969, Rusk noted that the "China" question was one of the central concerns of the bombing of North Vietnam and indeed was indirectly a cause of the *gradual* escalation of the war which so infuriated the U.S. military and the "hawks" of public opinion.[81]

Harry McPherson, who served in a number of posts under Kennedy and Johnson including special counsel, said in a 1969 interview that a military briefer had once admitted, "'To win the war in the classic sense of win would require an invasion of North Viet Nam or Laos or Cambodia or all three, to cut off the enemy's supplies.' But that everyone knew that the political inhibition of 'the presence of China' would not allow it."[82] In another interview, McPherson asserted that Johnson understood that

> the only way to really win the war was to invade North Viet Nam. There wasn't any other way. We thought we could do it by bombing them for a time, but I think he kind of gave up on that early although he never said so. I never heard him say so. I don't believe he changed his belief that he could do it bombing. But the only way to do it was to invade. But if you did invade, you'd risk an awful big war with China and Russia, so you couldn't.[83]

Defense Secretary Robert McNamara, the arch-architect of Johnson's Vietnam War, recalled in a 1975 interview, "Let me simply say here that in one very important respect our policy in respect to Vietnam succeeded. We avoided a military confrontation with either the Soviet Union or the People's Republic of China. This was a major objective of the president from the beginning. It was certainly a major objective of mine."[84] Rusk admitted the problems that the China-avoidance strategy caused but defended it by arguing, "Our problem [was] to prevent World War III. I said we learned the lessons of World War II, but no one is going to learn any lessons from World War III."[85]

Many critics noted the split-personality war that Sinophobia produced: a "limited" war that dragged on with no hope of actual victory, a conflict earning Johnson the rage of both the hawks and doves. In body bags, time, and spending it was the "seven years War" that George Marshall had pointed out long before that democracies cannot fight successfully. In his stinging critique of LBJ's war, military historian Harry Summers cites the failure to extend the land war into North Vietnam as the crucial factor ensuring an eventual Communist victory. As he puts it, "To return again to General Jackson's advice, our error was not that we were fearful of the dangers of nuclear war and of Chinese or Russian intervention in Vietnam. These were proper concerns of the military strategist. The error was that we took counsel of these fears and in so doing paralyzed our strategic thinking."[86]

Ellsworth Bunker, who served as ambassador to South Vietnam from 1967 to 1973, later complained that the United States did not cut the Communist supply trials in Laos because "there was fear it might provoke the Chinese, but [it was unwarranted because] we could make it very clear to the Chinese we had no designs on China or on North Vietnam. North Vietnamese could have any kind of government they wanted as far as we were concerned."[87] Journalist Peter Braestrup, who was from 1960 to 1968 the Saigon bureau chief of the *Washington Post*, in a later interview made a cogent analysis of the dilemma such a paradigm created: "Fear of China seemed to be a 'phobia' of Johnson: 'If I go in there then it will be like Korea and the Chinese will come in and I'll have a little disaster.' I mean he had all these phobias. Yet the CIA was telling him that the Chinese were too screwed up by the Cultural Revolution, at least in 1966, so that you don't need to worry about China. China's not going to do anything unless you push them to it, right?"[88]

F. Edward Hebert, a Democratic congressman from Louisiana, in a 1969 interview commented that he once grilled Secretary of Defense Robert McNamara "why a nation as mighty and as powerful and as resourceful as America can't defeat a little five-and-ten-store nation like Vietnam?" Hebert was outraged when McNamara's answer was that he lacked troop strength. Hebert noted, "They're always scared of China, scared of Russia, scared of China, scared of Russia. We shouldn't have gone in. When you pull a gun, you be ready to pull that trigger. If you aren't ready to pull that trigger, don't you pull that gun. Make up your mind then."[89] Soldiers in the field were no less bitter. In his memoirs, Army Intelligence Major Rene Defourneaux, who was part of a special team that first contacted and *trained* Ho Chi Minh guerrillas in 1945 and then worked with friendly Meo tribesmen during the Vietnam War, expresses his bitterness as follows:

> We had taken on France's obligations in Southeast Asia, but did not know how to carry them out. Two of our presidents had given the wrong message to the North Vietnamese. We violated the international rules of warfare by fighting first and then applying sanctions later. We refused to help our friends and made it easy for our opponents to manipulate not only U.S. public opinion, but that of that of the world. Winning was no longer part of our vocabulary. . . . No one wanted to hear the truth, no one seemed to care.[90]

Today many similar stories and feelings of resentment are generated from veterans of Vietnam. Such sentiments were behind the numbers of a 1980 poll conducted for the Veterans Administration, where 82 percent of veterans who were classified as "heavy combatants" in Vietnam thought that "the war was lost because they were not allowed to win." Historian and journalist Stanley Karnow adds, describing this result, "astonishingly, 66 percent indicated a willingness to fight again, presumably under fewer limitations."[91]

Although it was unconfirmed to either Johnson or his critics, the basic premise of Sinophobia was correct. Mao had chosen a hard line on Indochina and had made plans to intervene if America invaded North Vietnam.[92] Yet, the argument of LBJ's detractors on the left and right was also correct: fighting a war under the contradictions Sinophobia imposed was absurd. Dean Rusk, in a defense of the administration's polices, noted, "So long as I was Secretary of State there was never an initiative from Hanoi that could be described as a peace-feeler. The initiatives always came from somebody else, either ourselves or third parties. I cannot recall a single instance in which there was an initiative from Hanoi that could be described as a peace-feeler."[93] We now know why, its essential survival guaranteed by China, the Hanoi regime did not fear America enough to accept any substitute for complete victory.

To reach out to China, Johnson made some moves unthinkable in the Dulles era. At one of the Warsaw meetings in August 1964, the United States explained, after the Gulf of Tonkin incident and the escalation of U.S. involvement in Indochina, that any bombing of North Vietnam was not intended as an attack on the PRC.[94] On July 12, Johnson made a public statement calling for "free flow of ideas and people and goods" to China.[95] In December 1965, the travel ban was lifted for medical personnel traveling to China for "professional responsibilities." On January 9, 1966, scholars and writers were exempted from the travel ban as well. In February Johnson commented that America needed to find a way to establish normal relations with China.[96] On June 26, 1967, President Johnson met with Romanian prime minister Ion Gheorghe Maurer and passed on to him a number of assurances about China, with which country the Romanians had better relations than many other Communist nations. Johnson mentioned that he would "work out ground rules so that we can avoid nuclear war."[97]

But no progress was made because, in Washington's perception, and probably in fact, China was shutting down its diplomacy rather than opening up. In response to the Johnson Romanian outreach, for example, according to the State Department notation: "The Embassy in Bucharest reported . . . that Maurer . . . found the Chinese preoccupied with the idea the United States was preparing to attack the China mainland and had told them what Johnson had said to him, but that they were not convinced."[98] Walt Rostow commented that for Johnson, China "was not something we could do anything about."[99]

Yet, in the last weeks of Johnson's tenure, Walt Rostow did suggest that Johnson undertake some small measures of loosening trade restrictions for the sake of "a modest response to the faint signals from Beijing suggesting possible changes in their position. Although we would describe the change as merely a technical adjustment in our regulations designed to remove frictions with our allies, Beijing might interpret it as a sign of flexibility in our own position."[100] Secretary Rusk vetoed Rostow's suggestion of ameliorating trade relations even slightly, saying, "It should be done by someone else."[101] It was.

4. Pictures: China in the Mist

Socially, culturally, and politically, many who lived during the early 1960s, or who have commented on the period afterward, noted that at least through 1965 and possibly even 1966 many aspects of the era closely resembled the 1950s. The middle and late 1960s, however, typify the great upheaval that we now call "the '60s." Perhaps not completely by coincidence, Time*'s editorial stance toward China shifted in 1967, when Henry Luce stepped down from full editorial control. During the early '60s, China coverage in* Time *was simply a continuation of the previous period. The Chinese leaders are visualized as monsters (typically evil dragons) and tyrants, and government policies are described and shown as resulting in untold human miseries. But as the decade proceeded, cracks in the black-and-white division of the world began to appear. First,* Time *noted—as had become obvious—the growing split between China and the Soviet Union, which eventually led to outright military conflict on a small scale in the Far East. Second, the China Lobby—the stalwart defenders of Taiwan, of which Luce was the patron—began to fade in power. A new generation of Americans and politicians, who held none of the animosities, prejudices, or painful memories of their forebears, began to have more mixed feelings about Sino-American relations. A series of presidents and influential politicians and advocacy groups, at first mutedly and then more loudly, started to call for better relations and even normalization. By the late 1960s and the election of Richard Nixon, the geopolitics of the world had changed, and a Cold Warrior president would begin the intricate steps that would lead eventually to one of the most famous handshakes in diplomatic history. Most remarkably,* Time *magazine, along with almost all the American press, would accept, if not actually welcome, rapprochement. The images of evil, brutality, and mass murder were forgotten, and while the dragon did not become a panda, the view of China in the American press in general, and in the most China-focused American news publication in particular, would radically alter.*

The overall spread of China pictures by year from 1960 to 1967 in *Time* is shown in table 4.1 and constitutes the "mist" years of China coverage—the lowest points in the years I examined. China is not only little seen, but what we do see—as shown in the assessment of social dysfunction—is highly negative, a dangerous, uncontrolled, volatile, tyrannical China as much as any scene from the Korean War.

Looking at individual pictures from 1960 to 1967, we see some hints of change in the late 1960s—especially in contrast to the rest of the decade. *Time's*

Table 4.1. China Pictures by Year, 1960–1967

Year	*Freq.*	*Percent*
1960	23	1.2
1961	31	1.7
1962	33	1.8
1963	28	1.5
1964	31	1.7
1965	22	1.2
1966	56	3.o
1967	49	2.7

Table 4.2. *Time* China Covers by Year, 1960–1967

Year	*Covers*
1960	1
1961	0
1962	1
1963	1
1964	1
1965	1
1966	0
1967	1

Table 4.3. Surface Area of China Pictures by Year, 1960–1967*

Year	*Freq.*	*Percent*
1960	465	1.0
1961	1,029	2.2
1962	746	1.6
1963	746	1.6
1964	735	1.6
1965	492	1.1
1966	1,255	2.8
1967	799	1.8

*This table shows the yearly surface area of China pictures in *Time*. The calculation of surface area is based on:
Two pages = 187 square inches.
One page = 93.5 square inches.
Half page = 46.75 square inches.
Quarter page = 23.38 (rounded) square inches.
One-ninth page = 10.39 (rounded) square inches.

Table 4.4. Social Dysfunction in the PRC by Year, 1960–1967*

Year	*Percent*
1960	47.8
1961	80.6
1962	78.8
1963	82.1
1964	58.1
1965	63.6
1966	58.9
1967	95.9

*This table shows the percentage of pictures of social dysfunction in the PRC in *Time* by year in relation to the total number of pictures of the PRC.

coverage of China after the passing of the Taiwan Straits crises could be described as light but hostile. For example, a March 21, 1960, picture of a "Red China militia girl" shows one of several young ladies wearing a padded uniform pointing a machine gun at the camera. Her expression is more blank than grim. The article derides China's new feminine aesthetic, mocking the "emancipated women of Red China [who cry], 'what was beautiful is now ugly; what was ugly is now beautiful.'" The event described was International Women's Day, but *Time's* opinion of these "eager women of the militia" is disapproving. The women are, as seen by the publication, obsessed with power, ready to denounce members of their own families; they scream joyfully at executions, and toil in communes "as if possessed." But "the weirdest sight of all," according to *Time*, was "long lines of dutiful men who had been sent out by their women before dawn to wait, shopping baskets in hand, for the markets to open."

Other stories in *Time* in the early and middle part of 1960 deal speculatively with events in Formosa, such as who would be president after the tenure of Chiang Kai-shek, and a memorial to Claire Chennault, illustrated by a photograph of his Chinese wife and Madame Chiang. *Time* reserves its glee for the negative welcome of Zhou Enlai in India in light of border tensions between the two countries. *Time's* point is made via an Indian newspaper, which is quoted as saying, "China does not come here as a friend; she comes as an aggressor." *Time* tells the reader that India was "preparing for trouble," and one gets the impression that the newsmagazine was cheering for it. Other stories and pictures of the year deal with refugees from the PRC in Hong Kong and Macao; surveys of the specific defense perimeter of the United States; and the contrasting military pools of the Communist bloc and Western allied countries, on land and on sea.

The August 6 issue of *Time* displays a cover in which China is shown for the first time in the background rather than as a foreground menace. A smirking, beret-wearing Che Guevara is featured in a painting; over each of his shoulders,

Mao and Khrushchev scowl. Other *Time* stories about China detail literary repression in the country but also note, as an August 22 piece puts it, "After almost eleven years of Communist rule, China has gained not at all in the desperate race between food production and population increase." A picture of young Chinese women hoeing a dirt field is identified as "commune workers in northeast China." The subcaption, "Such an infinitely beautiful future," refers sardonically to Khrushchev's reported castigation of the commune system as a faulty way to achieve the paradise of Communism.

Other stories in the later part of the summer deal with religious persecution in the PRC. One exception to the almost uniformly positive Taiwan coverage is the picture in the September 19 issue of a democratic dissident in Formosa who was critiquing Chiang's one-party rule. This kind of story appeared in *Time* throughout the 1950s and '60s; it is frankly unfair, then, to pretend, as some critics of Luce have done, that *Time*'s world was all black and white. Nevertheless, the criticism of Formosa never reaches the level of absolute censure given the People's Republic of China. Chiang's regime is that of a friend who stumbles and who should be occasionally tweaked to right himself—that is, to become more like America. To cap this point for the year, as the U.S. presidential election entered its last days, *Time*'s October 24 issue shows a map of Formosa, the Fujian coast, and the offshore islands in relation to the debate between Kennedy and Nixon, in which the fate of Jinmen and Mazu was discussed. Overall, Nixon's position is supported by *Time*, with that candidate being allowed extensive quotes and factual support for them. Many other authorities are marshaled to counter Kennedy's allegations that Jinmen and Mazu are not worth fighting for.

News stories and pictures continue the negative trend. A cartoon in the February 3, 1961, issue shows a diminutive Khrushchev and Mao facing gaunt civilians over an empty bag labeled "crop shortages"; the accompanying story details the failure of Chinese and Soviet agricultural programs (see figure 4.1).

A picture on March 24 shows "Chinese commune workers digging irrigation canal," which accompanies a story on "the great leap backwards." This would be the theme of a number of China pictures during the year and indeed for the decade. For example, in the issue of October 16, with an article on "Marxism fails on the farm," there is a picture of a line of hoe-wielding "Red Chinese workers on agricultural commune." The subcaption comments, "In the factory without walls, starvation rations." A similar story, but without illustration, is found in the October 27 *Time.*

Other stories focus on battles in Indochina, especially in Laos; these are illustrated mostly with maps, which depict Red China as the source of or as a co-conspirator in infiltration by Communist guerrillas. The Sino-Soviet split, however, first receives prominent treatment in *Time*'s July 14 issue. A series of cartoons from newspapers shows Mao and Khrushchev fencing, hammer versus

"Nature And Some Minor Officials Have Failed You"

---copyright 1961 by Herblock in The Washington Post

Figure 4.1. February 3, 1961: Cartoon, "Nature and Some Minor Officials Have Failed You." (Copyright 1961 by Herblock.)

sickle, as rival fisherman vie for a fat fish labeled "Red world leadership" and two distance runners chafe at being handcuffed together. The accompanying story provides the long list of verbal attacks made by each side. Interestingly, Mao's China is repeatedly pictured as the aggressor, especially in relation to attacking the West. *Time* still has its doubts, wondering whether the reports of Khrushchev's moderation are just propaganda to win more friends in the West. Later, in *Time's* November 3, 1961, issue, although a picture shows Zhou and Khrushchev shaking hands, an accompanying story explains the "long quarrel" between the two nations. The Sino-Soviet split is mentioned in several other instances throughout the year.

An interesting development in Sino-American relations is reported in the August 11 issue. Chen Cheng, the vice president and premier of Nationalist China, is shown smiling with President Kennedy in a picture subcaptioned, "A clearer, more hopeful view." The story details some of the disagreements that the ROC had with the incoming administration and fears that a radical change in China policy was imminent. As related in the historical background above, Kennedy had by this time decided not to budge on China and accepted Eisenhower's advice-cum-warning. With a mild chiding at Kennedy's earlier ambiguity, the story informs the reader, "At last aware of the need to reassure a good friend, the administration spared no effort to make sure that Chen went home with the right ideas about U.S. policy." Although disagreements continued, "thanks to John Kennedy's firm statement that the U.S. view of Red China has not changed," Taiwan was much more pleased about the situation.

China's agricultural disaster—which was, as we have seen, just as real and terrible as *Time* made it out to be—is the entire feature of the newsmagazine's December 1, 1961, issue. The cover shows a gray, stubble-headed old Chinese man, head bowed in what we must interpret as despair. He is identified as "Red China's economic planner, Li Fu-Chun." The banner headline reads, "Behind the China debate, hunger, and hate." The debate is over China's admission to the UN, but *Time's* editorial stance is to shift the issue to China's internal problems. These are symbolized in the cover picture of a landscape of dry, cracked earth, an overturned and empty rice bowl, and a pair of chopsticks, one of which is broken, we presume in frustration by its famished owner. In the internal description of the major story, another image shows "Chinese refugees arriv[ing] in Macao." The letter from the publisher reads, "Covering Communist China is the toughest problem in journalism today. It is foolish to ignore one fourth of the world's peoples; but it is impossible to report the news out of China with any of the accustomed safeguards of trained eyewitnesses seeing for themselves and able to verify what they are told. Beijing allows only two permanent Western correspondents . . . and it shuns their requests and restricts their travel." The note then goes on to describe other sources of learning about what is going on in

China, mostly based in Hong Kong, such as refugees, Jesuit priests, and "a small colony of experts" from other countries.

Inside the issue, the pictures are varied. Those that might be considered positive, those from Eastfoto, are oppositionally captioned. A huge parade at Tiananmen Square that includes a statue of Mao (credited to Eastfoto) is subcaptioned: "On a diet of tree bark and rotten fish, a panoply of messianic power." In another image there are several not-so-prosperous-looking boys reading comic books. This picture could have been taken on any street corner or apartment vestibule in Asia, certainly in South Korea or Formosa. But here, they are labeled "indoctrinated children [who] pore over comic books which tell sublime achievements of Red heroes in war, industry, sports." More negative still is an image of several naked and half-naked children standing in line to lick an ice block. The caption tells us that these are "thirsty children on a hot day." Then we are informed, "refrigerators are rarities." *Time*'s slant is apparent here as well: refrigerators are rarities at this time in western Europe, all of Asia, Latin America, and not a few parts of the United States. A larger picture shows several children, one of them naked, picking through or inspecting "coal substitutes dried in a Beijing side street." The caption tells us further that these are "scavenging children" who collect coal dust from the detritus of "poor-quality Chinese coal" to refashion it into "crude briquets." Another image shows a scrawny old man picking at a dirt plot. This is a "tiny private plot two feet wide . . . worked by [a] Hopeh peasant. He may sell its produce, but furrowed field belongs to commune." The caption of a picture of an "urban commune" of blank- or grim-faced women sitting in the decrepit entrance to a building informs us that these communities "proved even greater failure[s] than rural ones."

Even pictures that bespeak some level of prosperity are editorialized as demonstrating the failure of Red China's agricultural revolution. Some female peasants work in a verdant rice field, but we are told that they are using "homemade sprayers and homemade insecticides to rid fields of parasitic pests." Elsewhere, we see "girl water carriers in Hunan province, home of Mao Zedong [and] soldiers in [the] grim fight against parched earth." A tiny roadside market is labeled "free." We are told that this is where "peasants can sell [a] small part of their produce." Another image shows the grimy, littered backyard of an "average family." Amid the garbage that passes for furniture, a child squats on filthy brick while a mother works over a homemade-looking stove. These people, *Time* tells us, live "in [a] warren less than a hundred yards from Beijing's elegant embassy row. Has only outdoor stove but is still relatively well-off. On hot nights, city Chinese move bedding to streets rather than stay in crowded, closed rooms." In contrast, a "Red elite family" lives in a "model flat containing two rooms and kitchen."

Another picture of men pulling a two-wheeled cart laden with metals, wires, and machinery tells us that they move "warily forward on two legs." A cartoon strikes home the unambiguous message of *Time*; again, it is one borrowed from a foreign publication (see figure 4.2).

A gaggle of skeleton-like Chinese peasants reaches out hands and empty rice bowls to a porcine Mao; he lectures them with finger pointed, holding an illustration of a bomb and a mushroom cloud. The caption reads, "But we'll soon have a wonderful A-bomb." In sum, *Time* does not solve "the toughest problem in journalism" through photography, because it resorts to question-begging captions that distort the metonymic claims they make about the pictures. Every scene here that proves the agricultural disaster in China—which really did occur—could be punctured by a critic of *Time* or a defender of Mao immediately. Certainly, no one who had traveled elsewhere in Asia—or in rural or urban slums in America—could look at pictures like these and say that they were unique to Communist China.

Time's coverage of China in 1962 is thin. Some stories deal with continuing tension with India, the fall from grace of Marshall Peng Dehuai, and a survey of America's foreign aid commitments throughout the world, including Formosa.

Figure 4.2. December 1, 1961: Cartoon, "But We'll Soon Have a Wonderful A-Bomb." (Reprinted from Fritz Behrendt—Algemeen Handsblad.)

Time's April 6 issue, however, is a focus on the border areas of Red China in south Asia, including the Himalayas, Kashmir, and Bhutan. There are some negative references to China having "made heavy inroads" into some of the regions. The fighting in the Himalayas and the Chinese occupation is detailed, but there are no images of this "struggle for the roof of the world," that is, the Himalayas, "where India is struggling to hold back Red China's hordes." *Time*, while in the past expressing hopes that India's Nehru would stand up to China, here is caustic at the "curiously torpid" behavior of Nehru and his associates. *Time* lectures, "India tried to buy Red China off by championing its admission to the United Nations [and] opposed all U.N. attempts to condemn the Chinese for their conquest of Tibet. The feeble Indian good-neighbor policy only encouraged the Chinese to look southward with greater interest."

China coverage continues with more stories about agricultural disasters and food shortages, and the escalating war in South Vietnam and Indochina. China is pictured in all the maps of this conflict, which is a major difference from the portrayal of the relationships in the Korean War. Another focus of China's problems is the stream of refugees attempting to escape. Formosa, for the first time in the year, comes under scrutiny with reports of a massing of Chinese troops across the straits in Fujian. *Time*, however, questions whether the crisis was of use to Red China wanting to divert the world from its farm crisis and political problems and to the Kennedy administration, for which the predicament "provided a front-page diversion from the troubled economy." The story also discusses yet more wars of words in the increasing Sino-Soviet tension. A photograph shows a "Jinmen sentry facing Red China." A soldier on a small hilltop overlooks the clear sea on the two miles of Amoy Bay to the hazy mainland. The subcaption reads, "Something of a grass menace on both sides." Other stories throughout the year also focus on China-Indian tensions, refugees, and China's internal troubles. Even a story on the wives of the major Communist leaders makes an allusion to "China's woeful economic failures."

The issue of November 2 shows in maps the "war on top of the world." *Time*, interestingly, covers the Red Chinese attack reasonably dispassionately, saving its best caustic editorializing for Nehru, who is described as a man who, instead of standing tough, "pathetically harped on Red China's 'lack of gratitude' for India's speedy recognition of Beijing and conceded that the invasion has 'made us realize that we were getting out of touch with reality in the modern world and living in an artificial atmosphere.'" In short, *Time* is crowing victory that Nehru, whom the magazine has been describing as naïve and practically a dupe for Red China, has finally been paid back in full by the nation he appeased.

The Sino-Soviet split, in pictures and cartoons, is covered in several following issues of *Time*, and then, on November 30, a cover picture delivers the capstone blow in the newsmagazine's castigation of Nehru. The Indian prime minister is featured on the cover in a painting; he wears his white jacket and cap; his eyes are downcast, his head and shoulders stooped, with an expression of preoccupation and perhaps disappointment. The banner headline reads, "India's Lost Illusions." Dominating the background of the image is the gigantic head of a flush-faced, scowling, contemptuous Mao Zedong. Maps within the issue display China's military victories in the border areas; a photograph from 1957 of Nehru and Zhou hugging reminds us of how Nehru arrived at this point of despair.

Time's coverage of the China-India war continues into 1963. In the January 11 issue, there is a brief story featuring a picture of some "Chinese frogmen"; the piece discusses raids and counterraids along the Chinese coast between the Communists and the Nationalists. Other coverage includes the Sino-Soviet split, with a table showing the differences between Mao and Khrushchev on major issues; the continuing guerrilla wars in Indochina; and internal problems in the PRC. But the main focus is the estrangement between the old Communist allies, with pictures of conferences and leaders interspersed with insults and attacks exchanged between the two groups.

The Chinese dragon returns on the cover issue of September 13, 1963 (see figure 4.3). Now it is broken toothed, wide eyed, furious, but somehow less threatening, because it is transformed from an autonomous creature into the masthead of—here the metaphors mix wildly—a Viking longship. The vessel itself is overflowing with thousands of faceless people who variously spill off the sides, try to row in confusion, hold up propaganda posters, pictures, and flags, and even prop up a gigantic missile warhead (a reference to the PRC nuclear weapons program). The boat, however, is bursting at the seams: planks have sprung open; steel patches appear on the hull; even the masthead itself appears to have broken off and then been nailed back into place. On top of the dragon's head are the various Chinese leaders exhorting their "masses." Mao holds flowers and declaims with one outstretched hand, more like a Don Juan than a dictator. Zhou glowers at the throng below and holds a whip. The other Chinese leaders exhort or plead. The banner headline reads, "Red China: The Arrogant Outcast." Inside the magazine, as in the previous year's China-cover issue, the focus is on China's problems. In one cartoon, a smiling Mao at his desk informs a black man, "Of course there is no discrimination in China. Here we enslave everybody." The pictures and captions are less caustic. Interestingly, *Time* ends the year with perhaps the only unambiguously positive story about Communist China for almost a dozen years: a medical triumph by Chinese surgeons who reattached the severed hand of a factory worker. The story admits, "The Shanghai team achieved more success than any similar American effort."

Time begins its coverage of China in 1964 with a cartoon ridiculing Charles de Gaulle as a Chinese emperor complete with umbrella and Buddha-like belly.

Figure 4.3. September 13, 1963: Time cover painting, "Red China: The Arrogant Outcast." (Artist: Artzybasheff. Copyright 1963 by Time, Inc. Reprinted with permission.)

The article concerns the French leader's plan to "soon recognize Communist China, a move prompted by the faint hope of reviving French influence in the Far East." American and West German officials are quoted condemning the move, but *Time* also offers quotations from leaders or politicians in former French colonies. The issue at hand: whether the defection from the ranks of the West might prompt a recognition movement from the People's Republic of China. Other China-related images over the next few months detail Zhou Enlai's trips around the world. For the first time, on March 13, comes the face of a Chinese leader who is soon to go into purged eclipse but who will return in the 1980s to be the most familiar representative of China. A Romanian delegation is pictured while visiting Beijing; at the meeting table is one squat figure holding a cigarette. His name, Deng Xiaoping, is not noted in the caption or the body text. The story itself continues the theme of the failure of the Chinese economy, mentioning "Red China's desperate need for refining equipment."

While no entirely new chapter of Sino-American relations begins in the spring of 1964, for the first time Americans are taught its basics. An April 3 map of Southeast Asia displays all its nations with dominoes, some of which are about to topple. The map is labeled "Asian Dominoes" (see figure 4.4).

The reference, of course, is to the domino theory, which first flourished at the time of the 1954 Geneva talks. The visual and political lesson: if one small Asian country falls to Communism, the next will topple as well. The story emphasizes a new fulcrum for American policy in east Asia: a country called South Vietnam. A "veteran U.S. foreign officer in Hong Kong" is quoted as saying, "Southeast Asia is the Spain of the 1960s. If we can't and don't win here, how can any friend of ours believe we can win anywhere?" The story itself asserts, "The key to the situation remains the U.S. struggle to keep South Vietnam from falling to Communism." The image is a striking visual illustration of how the case for war—what would actually become the wider wars in Indochina—was made for the reader. No picture perhaps better sums up how a complex policy debate, or what should have been a complex policy debate, was "Lucized" into a simple call to arms.

The Sino-Soviet split is now a regular feature of any mention of Communist-bloc doings. Almost always, *Time*, while not holding much sympathy for Khrushchev, casts Mao as the villain of the divorce. The issue of April 24 contains a retrospective history of "Communist isms and schisms" with a face-filled timeline. The focus is on the Sino-Soviet split, with critical quotes coming from Khrushchev and Mao of each other and also of unnamed sources. One of the pictures that illustrates this divide is that of a smiling Mao and Khrushchev on a podium facing each other, in tense standoff. Other pictures of the spring focus on Maoist guerrillas in Nepal and the growing crisis in Indochina.

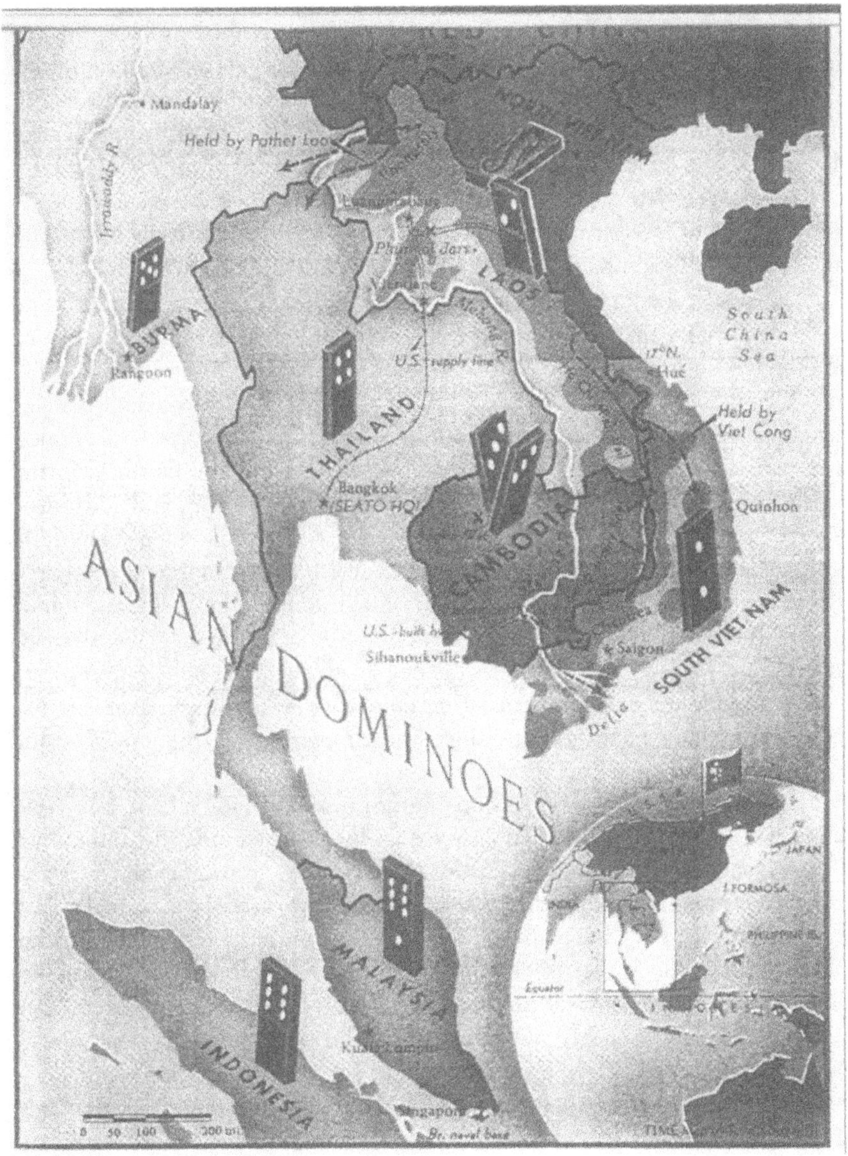

Figure 4.4. *April 3, 1964:* Time map, "Asian Dominoes." (Copyright 1964 by Time, Inc. Reprinted with permission.)

Time's April 17 issue displays one of the other perennial problems of the magazine. We recall that earlier, *Time* had admitted that Formosa's army in the early 1950s was much larger on paper than its effective strength for battle, a difference of hundreds of thousands of troops. Yet now, in memorializing Douglas MacArthur, who had died the week before the issue went to print, *Time* recounts in a map "Mac's victory plan," which was premised on those same faulty numbers. We note of interest that northwest China is still labeled "Manchuria," although now *Time* admits that the region contains "Red Chinese air bases."

In a map in the June 12 issue, North Vietnam is colored as a "Chinese puppet." *Time* argues that the policy options available to the United States—and here the bells of retrospective history ring sad and long—do not include withdrawal from South Vietnam, because "a U.S. pullout would convince all of Asia, including Japan, that Americans cannot be relied on as allies against Communism." *Time's* recommendation, made implicitly, is to "ignore North Vietnam for the moment, drop the U.S. 'adviser' role in South Vietnam, and add thirty to forty thousand U.S. troops to the fray. Some feel that this added strength in fighting men, plus the addition of perhaps a few hundred thousand newly recruited Vietnamese soldiers . . . would be enough to back the Viet Cong into a corner." Interestingly, *Time* does not seem to advocate the Korean War strategy of bombing North Vietnam since, as the newsmagazine notes now in retrospect, it did not work in Korea. Nor does *Time* note what exactly in geographic terms is the "corner" to which they refer since in their own very map South Vietnam is permeated with guerrilla strongholds.

The weeks to follow also focus on Southeast Asia, including Laos, but *Time* pauses in the July 31 issue to publish yet another piece on life in Communist China. The illustrations are from Eastfoto and include street scenes and a restaurant dining room. The story is very negative. In *Time's* attempt at some level of fairness, however, the Chinese people are praised, as in the honesty of guests who "never bother to lock their rooms." (This theme will reappear in coverage of China masses during Nixon's trip in 1972.) *Time* further notes, however, "Despite China's many wonders, Western visitors find the atmosphere depressing." On August 14, 1964, *Time* reports the fateful Tonkin Gulf attack on U.S. warships, which led to increased war powers for LBJ. In the map that shows the "area of action," China looms at the north; its army, naval, and air force strengths are provided.

In 1965, the first year after Henry Luce's semiretirement, *Time* is placed in the difficult position of having to recognize the growing acceptance of Communist China and its influence throughout the world. *Time* reports the business interests reasonably factually, but cannot help but deliver some negative editorializing. A picture of "Red Chinese buyers in Britain" is subcaptioned, "And a thirst for Commie Cola." The main coverage of China throughout the year is ancillary:

several maps of Indochina, with "Red China" pictured in slices of the north. But then, in the February 26 issue, China's dragon reappears on the cover of *Time.* In sunglasses and wearing a natty business suit, "Red China's foreign minister Chen Yi" is painted. This time, the dragon is not so much in the background as floating above Chen's head. It is a dynastic bronze, with large pointed ears, a huge nose, bulbous eyes, and barracuda-like serrated teeth. Its four feet, however, are hardly threatening claws. The banner headline reads, "The Enemy in Asia." The focus in the accompanying story is on "Pacific power," on display on a map listing the strength of forces of China and the United States in Asia. Chen, who is described as "the hatchet man," is shown in a number of friendly diplomatic photos, but the focus of the article is on his police state repressions.

Time's editorial policy on China seems to take a shift in this issue and it is an important one, especially as it occurs in the last two years of Henry Luce's life, when his direct grip on his publications was slipping. The unsigned story comments:

> For fifteen years the principal U.S. stand toward China has been not to recognize its Red regime. While there is no reason to change that stand at present, nonrecognition is no substitute for a policy. Since Korea, the U.S. in fact has developed not a China policy but a China mythology. One side of the myth holds that China is remorseless, implacable, omnipresent, and possessed of warriors who love nothing better than to die in "human sea" attacks. The other side holds that, like Russia and the satellites with whom the West has learned to live, Chinese Communists will in time grow softer, more reasonable. They may, although European Communism, superimposed on viable economies and workable political structures, is vastly different from the Asian variety. What is at stake in Asia is an undeveloped, politically shapeless region full of people who deserve better than the absence of chrysanthemums. The West certainly cannot impose capitalism or democracy on Asia with the air of a crusader. But it can work toward building free economies and free societies, even if socialist concessions have to be made. Malaysia and Thailand represent viable, hopeful alternatives to Tibet and Burma. In the meantime, the U.S. must hang on—and then hang on some more—in Southeast Asia. The operative word is patience, and essentially, patience is an Asian word.

This is an important statement, and perhaps marks *Time*'s entrance into a new era regarding China. The combativeness of the mid-1950s is gone. There is not even a suggestion of an uprising against Communism within China or of the Nationalists' retaking the mainland. It is also noted that the wall of containment may be militarily intact, in the Pacific at least, but it has collapsed in Indochina; moveover, the economic isolation of the PRC is no longer completely viable.

These are all statements that subsequent history would assess as factually correct. And even countries like Thailand, and less so Malaysia, have proven to be alternatives to either Communist autocracy or some corrupt Western-supported tyranny. Both of those nations went through those phases and, as of this writing, both have reasonably optimistic democratic futures. Thailand especially is yet another one of the smaller Asian "miracle tigers." But patience and hanging on, virtues in 1965, seem to be prophecies of doom when writing in the twenty-first century with the full tragedy of the Vietnam conflict available to our gaze. Lyndon Johnson did have patience; he did "hang on" in Vietnam. As we have seen, such a policy produced neither victory nor peace nor honor nor any benefit for America or the peoples of Indochina. *Time*, in other words, whereas before it shifted hard to the right on the issue of Communism in Asia, here is a voice of moderation, and from the perspective of 1965 the alternatives to moderation—a complete abandonment of Indochina to Communist takeover or a Goldwater-advised (MacArthur-inspired) all-out commitment to war in Southeast Asia—were as frightening to *Time* as they were to the American people and practically everyone in the U.S. administration.

As David Halberstam chronicles in his book *The Best and the Brightest*, and as noted in the historical background given above, the Johnson administration consistently committed fewer troops to Vietnam than the military felt it needed because it did not want to go "all out," and, of course, the great fear was of provoking China. As one historian put it, Johnson's men wanted to make escalation "as imperceptible as possible to the American public."[102] The result was what we call euphemistically "limited war," but, if we count the time from the first U.S. all-out commitment to the exit of the last major concentrations of U.S. troops from 1965 to 1972, we see the full scope of George Marshall's remark, that democracies cannot fight a seven-years war. Sometimes, policymakers, faced with no good options, choose the option of least resistance, even though in the end it will prove to be the worst option.

Throughout the year 1965, *Time* tends to focus mainly on the conflict in Southeast Asia. The repetition of China appearing in the map to the north of Vietnam is not just a geographic fact; it is a constant reminder of the possibility of China's entrance into the war. This is "Red China," not just Manchuria or some province. One wonders what would have been the effect if *Time* had followed the same policy in 1950. In fact, practically every depiction of China for the entire year of 1965 in every issue of *Time* is sliced into a map of Indochina. Visually, *Time* is reinforcing the administration's case that wariness of Chinese involvement should guide Vietnam policy.

Time's March 19 issue contains a cartoon that repeats an old visual metaphor: Mao Zedong, his traditional Chinese Mandarin hat flying through the air, fights a Brezhnev figure; Mao swings a sickle, Brezhnev a hammer. The story focuses on how, after Khrushchev's ouster, the Sino-Soviet split, far from healing, had become even more bitter. The metaphors are intriguing because earthy, elemental, backward China is pictured with the sickle while more industrial, advanced Russia holds the hammer; Brezhnev wears a czarist officer's uniform coat and a fur hat; Mao is bald and buck-toothed. *Time's* use of cartoons to illustrate the Sino-Soviet split is another sign of the lack of useful photography. There were at this time border clashes; no photographs came of them from Western publications. Alternatively, Chinese and Russian leaders were still meeting, and *Time* could always caption these photos with some version of "behind the smiles, hostility ferments." But the cartoon allows the full expression of *Time's* editorial stance, and this cartoon is produced in-house, not from a third-party newspaper.

Later in the year, Zhou Enlai is featured in one of his perennial Third World tours, with *Time* providing cutting captions and story content. In an image of a Communist Party conference in Bucharest, Nicolae Ceauşesu, the Romanian party leader, is pictured all smiles with Deng Xiaoping and Leonid Brezhnev. The story, however, details how Romania, like China, is breaking away from the Soviet orbit. *Time's* September 24 issue deals with—in connection to the Indian-Pakistan war—Chinese encroachment in the north of India. *Time* is still not forgiving of Red China. A picture of Mao and Zhou Enlai with Cambodia's Norodom Sihanouk is captioned, "How not to polish a nadir." The accompanying text reads: "Nominally, the occasion was the sixteenth anniversary of Red China's 'independence.'" *Time* here cannot help but put the word "independence" in quotes to place a question mark on the concept of a nation enslaved by Communism being "independent." The text resumes, "But Beijing's play for world headlines over the past three weeks had a subtler purpose than mere celebration. Beijing, which early this year seemed to have everything going its own way, last week stood at something of a nadir in the eyes of the world. A series of ham-handed debacles in Beijing's militant foreign policy has set Chinese propagandists to a frantic bit of image polishing abroad." The article goes on to list the "gaffes and circuses," including China's flare-up with India on its border, its failure to help North Vietnam, and a number of other foreign policy snafus.

The year1966, when the Johnson administration first seriously considered changes in China policy, began in *Time* magazine with yet another focus on the exact reason why Johnson felt some need to reach out to China. A series of maps displays the battle areas of Indochina. Two railroads connecting into North Vietnam and Hanoi are labeled "Red Chinese supply lines." One explosion marks "closest bombing to Red China." An "X" on China marks the spot of the country's first atomic test. In short, China is in the thick of the war in Vietnam. A cartoon in *Time's* January 14 issue shows an angel holding a sign asking, "Peace, please!" and American, British, Chinese, and Vietnamese leaders each pointing to the others. The main portrayal of China throughout at these years continues to be through maps. An exception, however, comes when China itself is described in a February 25 story that shows several pleasant scenes from the PRC, including a street scene, the Great Wall, and not-very-unhappy "coolies at work on dam." The story, in contrast, focuses on how Red China is "frustrated and alone." The reader is informed, "Relations with Russia are steadily growing worse." China's interior situation and its international relations are described as giving "Mao & Co. . . . plenty to worry about": population explosions, artistic dissent, and failed foreign policy ventures. Indeed, the portrait that *Time* creates—despite many pictures of Chinese diplomats smiling with foreign leaders on state visits and foreign diplomats smiling with Chinese leaders in Beijing—is that China's foreign policy is a failure across the board.

This last negative criticism is factual, but only to a point, and that point was a disaster for the United States. China had undergone severe foreign policy blows in the early months of 1966. The pro-Chinese leader of Ghana suffered a coup at home while on a tour of the PRC; a Communist uprising in Indonesia was crushed with a great massacre of the ethnic Chinese community. But a cartoon, borrowed from a Dutch newspaper, overviews *Time*'s description of PRC policy failure. A heavily bandaged and beaten Mao limps with a cane, his ice-packed and swollen head encircled with cartoon stars. An angry but diminutive Russian bear chomps on his coattails. Among the injuries that China has sustained include a broken arm labeled "Indonesia" and a broken leg (the right leg, although it is unclear whether or not this is a political reference) labeled "Vietnam." The accompanying story, "Communists—Fight of the Tigers," details more of the Sino-Soviet split. The story does not explain why Vietnam is such a catastrophe to China. Here it may be *Time* magazine's hope rather than reality that it is China that suffers from the Vietnam War and not America.

In the weeks to follow, Chinese foreign policy failures in south Asia and Indonesia are detailed, but the maps continue to remind us that China is always present. The issue of May 13 shows a picture of Mao with a Cambodian delegation and speculates on rumors of his ill health. A special *Time* essay, accompanied by more than a dozen pictures, is featured in the May 20 issue. The opening paragraph of "what the U.S. knows about Red China" repeats a *Time* tactic employed since the early 1950s of stating that, while it is very difficult to know what is going on in China, through different sources it is possible to reveal much "about a nation of such implacable hostility and resolute secrecy." A photo essay shows the "Red China tourists see." Featured scenes are peasants working in fields, statues of Mao, and propaganda posters. *Time*'s wry contrarianism has become more subtle. The quotation marks and harsh oppositional captioning are gone, but the writing still emphasizes an ironic view of what is being pictured. For example, "Inevitable on any tour is a collective farm. Above, happy workers plant rice," and "Identified by red handkerchiefs, children of the Young Pioneer Corps . . . are marched past Martyrs Park. Carefully selected, good pioneers become good Communists." The story ends with a call to engagement with China. Prorapprochement academics like A. Doak Barnet and Lucian Pye are quoted as insisting that change in China is coming soon, and that the United States should be willing to explore how to bring about that change. The change in the Luce-press sans Luce is also evident here: in the past, any such inclination would have been heavily undercut.

Pictures for the rest of the year focus on China's foreign policy, especially in Indonesia, Eastern Europe, and Indochina. There is generally little coverage, for the obvious reason of not having access to it, of the Cultural Revolution; what there is tends to be mocking. For example, in the August 19 issue, young men and women screaming and holding up rifles are described as participating in "Red Chinese demonstrations in Canton." The subcaption reads, "Chaos is good for what ails you." Another picture of Mao and Lin Biao—who would a few years later be purged—in the August 26 issue is subcaptioned, "Unlikely to solve the problems." Lin Biao is also featured on the cover of *Time*'s September 9 issue. He smiles, although in the magazine's portrayal this comes off more as an evil leer. Behind him appear the Chinese characters for the number two (meaning his is second in line to Mao). The banner headline reads, "The Chinese Nightmare." Inside the issue is probably one of the most detailed descriptions *Time* has attempted of the Cultural Revolution. The reader is notified about the purging of all Western literature and music and the banning of "happiness." China is described as being criticized by practically all other Communist countries, even Castro's Cuba. A photograph shows "expelled nuns crossing the border to Hong Kong." China is fundamentally portrayed as a land of "rage and ambition churned up by the Cultural Revolution." Its foreign policy objectives are described as including "victory in Vietnam."

Cultural Revolution scenes, typically of screaming young people holding up posters or pictures of Mao, reappear in *Time*'s September 23 issue. The accompanying story speculates on whether the terror, confusion, and fighting in China reflect some split in the leadership. The story further maintains that Mao has "brought terror of a new and terrible sort to his hapless land." In a cartoon taken from the *Washington Post* in *Time*'s October 28 issue, Mao is shown on a mountaintop holding stone blocks shaped like the Ten Commandments, except that they are labeled "Mao's thought." He is absolutely alone. On other mountains, groups meet at treaty tables; they include the "Manila Conference," the "New Delhi Non-aligned Nations Conference," and the "Moscow Soviet-bloc Conference." China's isolation, however, does not negate its danger. In *Time*'s November 4 issue, the last China image of 1966 shows a map of the nation that accompanies a story on the first successful test of a Chinese nuclear missile.

Time begins its coverage of "China in chaos" as the cover story of its January 13, 1967, issue (see figure 4.5).

The feature image once again juxtaposes a Chinese personage with a morphic dragon. Chairman Mao, lips pursed, squinting, dyspeptic, scans the horizon, but, one senses, he is not looking at anything in particular. His head, like some disembodied bust, is surrounded by a snakelike dragon whose body consists of the instantly recognizable lengths and guard towers of the Great Wall. The dragon is not facing us, as was the case with every other one of *Time*'s previous Chinese dragons, but is scuttling away, about to close the gap with its own tail. The metaphor is striking and brilliant: the empty god of Mao has walled off the nation and walled out the world. The China monster is also, by implication, about to—like some serpent from a mythological tale—swallow itself. The story contains a series of uncredited photos and other photos credited to Eastfoto and UPI featuring images of "kindergarten children jeering 'U.S. imperialist'" in a poster; the "imperialist" is a caricatured head of a bald American with a bandage on his temple. He has a cruel grimace, but he looks as much frightened as fierce. He has an extremely long nose which, in Chinese iconography, was a common technique for caricaturizing Occidentals.[103] Another photograph shows "Red Guards polling barge on lake"; yet another displays "pro-Mao posters on Hangchow Buddha." The subcaptions undergird *Time*'s editorial line, which is only somewhat less acerbic than it was in the Luce era of direct control. The events pictured are called "madness and mindless." The story details incidents from the Cultural Revolution and tries to gauge the difference between the winners (or "heroes," which appellation is obviously used ironically) such as Mao himself and his wife, Jiang Qing, and the losers, who are listed as including Deng Xiaoping and Zhou Enlai. Practically the entire story is dedicated to describing the outrages and upheavals perpetrated by the Red Guards in Mao's name. Mao is depicted as wanting to "throw China back into an age of simplicity and Spartan evangelical purity that it had never really known."

In the weeks to follow, China images focus on the Cultural Revolution, especially on denunciations from posters, placards, and demonstrators. The scenes are heartwrenching to the Western observer: humiliated men forced to wear placards denouncing their own conduct; a China riven by youth revolt of the worst kind. In the February 10 issue, *Time* shows a pictures of "Red Guards laying siege to Soviet Embassy in Beijing." The newsmagazine calls it "a trick as old as tyranny" that Mao is transposing internal disaster onto an external enemy, except this time the target is an old ally, the Soviet Union. Chinese students are quoted as demanding, "Hang the bastard Brezhnev." Although Red Guard images and demonstrations—burnings in effigy, humiliations—are the subject of a number of pictures, by late March *Time* claims that the Cultural Revolution is "slowing down" and that Mao is now "muzzling the dragon." A story that features a pic-

Figure 4.5. January 13, 1967: Time *cover painting, "China in Chaos." (Artist: Giro. Copyright 1964 by Time, Inc. Reprinted with permission.)*

ture of Chiang Kai-shek in Taiwan is less bellicose than any previous *Time* report. The voices for retaking the mainland typically heard during past crises in the PRC are described as muted or few. "Nevertheless," the story tells us, "Chiang stands ready for any eventuality. The country's six-hundred-thousand-man army is well-trained and well-equipped, and Chiang keeps eighty thousand troops poised and battle-ready" on the offshore islands.

Time's focus then shifts to the Indochina war. A map in the March 17 issue shows the "bombing areas" in North Vietnam; of interest is the region of "restricted areas," which include the "Red China border." The issue of April 14 reports that Mao had "changed his mind" and was once more sending the Red Guards into the streets. The next image of the events does not come until more than a month later, on May 26. Beijing sympathizers are shown demonstrating in Hong Kong; other images in later issues focus on more civil unrest, including attacks on Chinese diplomats in India. The link between China and the Vietnam War is, as always, made continuously in maps, but a July 28 story details how the Chinese are trying to "egg on trouble in other countries," including Laos, Hong Kong, Burma, and Japan. Through these weeks, of course, all the images are, from the Chinese point of view, exactly what they want the world to see of them, and indeed as usual most of the photographs come from Eastfoto.

In an August 9 story, *Time* makes one more focused connection between China and the Vietnam War. It locates a MiG base for the North Vietnamese within China and shows its location on a map. Bloody street fighting is also shown in pictures between crazed-looking and axe-wielding pro-Mao demonstrators in London. The remainder of *Time*'s 1967 China coverage focuses on more dysfunctions: its increasingly bad relations with other Asian countries; attacks on Chinese people and institutions in Indonesia; and more scenes of the Red Guards in their various "mass" apparitions. This last point is worth expanding. The Red Guards appear in *Time* as perpetrators of chaos and hatred. They fit well into yet another variation of the hordes of China, or the yellow peril.

Notes

1. Foundations of Foreign Policy, 1969–1972. Department of State. *Foreign Relations of the United States/Nixon/I/*, ed. Louis J. Smith and David H. Herschler (Washington, D.C.: United States Government Printing office, 2003), www.state.gov/r/pa/ho/United States.

2. Noam Kochavi, *A Conflict Perpetuated: China Policy During the Kennedy Years* (Westport, Conn.: Praeger, 2002), 250.

3. Arthur A. Schlesinger, *A Thousand Days: John F. Kennedy in the White House* (Boston: Houghton Mifflin, 1965), 480.

4. Stephen E. Ambrose, *Eisenhower, the President,* vol. 2 (New York: Simon & Schuster, 1984), 99.

5. Robert J. Donovan, *Eisenhower: The Inside Story* (New York: Harper, 1956), 132.

6. Ross Koen, *The China Lobby in American Politics* (New York: Harper & Row, 1974), 131.

7. Karen Shaw Kerpen, "Voices in a Silence: American Organizations That Worked For Diplomatic Recognition of the People's Republic of China by the United States, 1945–1979" (Ph.D. diss., New York University, 1981), 147–48.

8. Lee Riley Powell, *J. W. Fulbright and America's Lost Crusade* (Little Rock, Ark.: Roe Publishers, 1984), 48–49.

9. William F. Buckley, Jr., "To China with Nixon: Is There a Road Back?" in *Execution Eve and Other Contemporary Ballads* (New York: G. P. Putnam's Sons, 1975), 31.

10. Andrew J. Nathan, "Re-Examining the Cold War: U.S.-China Diplomacy, 1954–1973," *China Review International* 10, no. 2 (2003): 436–40.

11. I. C. Ojha, "Sino-American Rapprochement: Why Wasn't It Predicted?" *China Report* 9, no. 3 (1973): 53–57.

12. M. S. Kapitsa, *On Different Latitudes—Diplomat's Notes* (Moscow: Kniga i biznes, 1996), 75–76. My thanks to Svetlana Kulivova for her translation.

13. Cited in Zhisui Li, *The Private Life of Chairman Mao,* trans. Tai Hung-Chao (New York: Random House, 1994), 565.

14. F. E. Rogers, "Sino-American Relations and the Vietnam War, 1964–1966," *China Quarterly* 66 (1976): 293–314.

15. T. Tsou, "Statesmanship and Scholarship," *World Politics* 26, no. 3 (1974): 428–51.

16. Cited in Thomas Griffith, *Harry and Teddy: The Turbulent Friendship of Press Lord Henry R. Luce and His Favorite Reporter, Theodore H. White* (New York: Random House, 1995), 278.

17. See overview in Jonathan Rosenberg, "Between 'Paper' and 'Real Tigers': Mao's View of Nuclear Weapons," in *Cold War Statesmen Confront the Bomb: Nuclear Diplomacy Since 1945,* ed. John Lewis Gaddis, Philip H. Gordon, Ernest R. May, and Jonathan Rosenberg (Oxford: Oxford University Press, 1999), 171–93.

18. Roger Hilsman, *To Move a Nation: The Politics of Foreign Policy in the Administration of John F. Kennedy* (Garden City, N.Y.: Doubleday, 1967), 339.

19. Paige E. Mulhollan, "Walt W. Rostow Oral History Interview I," transcript, March 21, 1969, Lyndon Baines Johnson Library, 128.83.78.10/johnson/archives.hom/oralhistory.hom/rostow/rostow1.pdf.

20. John M.Murphy, "Crafting the Kennedy Legacy," *Rhetoric & Public Affairs* 3, no. 4 (2002): 577–601.

21. Doris Kearns, *Lyndon Johnson and the American Dream* (New York: Harper & Row, 1976), 178.

22. Moya Ball, *Vietnam-on-the-Potomac* (Westport, Conn.: Praeger, 1992), 2.

23. James C. Thomson, "On the Making of U.S. China Policy, 1961–1969: A Study in Bureaucratic Politics," *China Quarterly* 50 (April–June 1972): 220–43.

24. John M. Murphy, "Crafting the Kennedy Legacy."

25. Arthur M. Schlesinger, *A Thousand Days* (Boston: Houghton Mifflin, 1965), 425.

26. Nigel Hamilton, *JFK: Reckless Youth* (New York: Random House, 1992), 787–88.

27. Joseph Alsop, "The Feud between Stilwell and Chiang," *Saturday Evening Post*, January 7, 1950, 16.

28. Congressional Record, January 29, 1949.

29. William A. Arnold, *Back When It All Began* (New York: Vantage, 1975), 14.

30. James McGregor Burns, *John Kennedy: A Political Profile* (New York: Avon, 1959), 90.

31. John F. Kennedy, "A Democrat Looks at Foreign Policy," *Foreign Affairs* 36, no. 1 (1957): 50.

32. Vice President Richard M. Nixon and Senator John F. Kennedy, second joint radio-television broadcast, NBC, October 7, 1960, www.jfklibrary.org/60-2nd.htm.

33. N. Khrushchev, *Khrushchev Remembers: The Last Testament* (London: Deutsch, 1974), 488–89.

34. Khrushchev, *Khrushchev Remembers*, 490–91.

35. Chester Bowles, "The China Problem Reconsidered," *Foreign Affairs* 38, no. 3 (1960): 476–86.

36. Adlai Stevenson, "Putting First Things First: A Democratic View," *Foreign Affairs* 38, no. 3 (1960): 191–208.

37. Kenneth T. Young, *Negotiating with the Chinese Communists: The United States Experience, 1953–1967* (New York: McGraw-Hill, 1968), 238–39, 253. See also Nancy S. Simon, "From the Chinese Civil War to the Shanghai Communiqué: Changing U.S. Perceptions of China as a Security Threat" (Ph.D. diss., Johns Hopkins University, 1982), 225; Gordon H. Chang, *Friends and Enemies: The United States, China, and the Soviet Union, 1948–1972* (Stanford, Calif.: Stanford University Press, 1990), chap. 8; James Fetzer, "Clinging to Containment: China Policy," in *Kennedy's Quest For Victory: American Foreign Policy, 1961–1963*, ed. Thomas G. Paterson (New York: Oxford University Press, 1989), 178–97; Rosemary Foot, *The Practice of Power: U.S. Relations with China Since 1949* (Oxford: Clarendon Press, 1995), 178.

38. Jasper Becker, *Hungry Ghosts: China's Secret Famine* (London: J. Murray, 1996).

39. Thomas P. Bernstein, "Stalinism, Famine, and Chinese Peasants: Grain Procurements During the Great Leap Forward," *Theory and Society* 13, no. 3 (1984): 339–77; Judith Bannister, "Population Policy and Trends in China, 1978-1983," *China Quarterly*, no. 100 (December 1984): 717–42.

40. Claudius Colombo, "Chinese Communist Perceptions of the Foreign Policy of John F. Kennedy, 1961–1963," (Ph.D. diss., New York University, 1982), 21.

41. Colombo, "Chinese Communist Perceptions," 25.

42. Colombo, "Chinese Communist Perceptions," 31.

43. Colombo, "Chinese Communist Perceptions," 37.

44. James Thomson, Jr., "On the Making of U.S. China Policy, 1961–1969: A Study in Bureaucratic Politics," *China Quarterly*, no. 50 (April–June 1972): 220–43.

45. United States Department of State, *Foreign Relations of the United States: Diplomatic Papers, 1961–1963*; vol. 22, *China; Korea; Japan*, www.state.gov/www/about_state/history/foreign relations of the United Statesxxii/01to50.html.

46. Roderick Macfarquhar, *Sino-American Relations, 1949–1971* (New York: Praeger, 1972), 182.

47. James Thomson, Jr., "On the Making of U.S. China Policy, 1961–1969: A Study in Bureaucratic Politics," *China Quarterly*, no. 50 (April–June 1972): 222–23.

48. Schlesinger, *A Thousand Days*, 483.

49. United States Department of State, *Diplomatic Papers, 1961–1963*, vol. 22, *China; Korea; Japan.*

50. CQ, *China and the U.S. Far East Policy*, 1945–1967 (Washington, D.C.: Congressional Quarterly Service, 1967): 127.

51. Stanley D. Bachrack, *The Committee of One Million: China Lobby Politics, 1953–1971* (New York: Columbia University Press, 1976), 20.

52. Ralph G. Martin, *Henry and Clare: An Intimate Portrait of the Luces* (New York: Basic Books), 379–83.

53. Schlesinger, *A Thousand Days*, 479.

54. Theodore C. Sorensen, *Kennedy* (New York: Harper & Row, 1965), 665.

55. Sorensen, *Kennedy*, 665–66.

56. Foster Rhea Dulles, *American Policy toward Communist China* (New York: Thomas Y. Crowell, 1979), 193.

57. Roger Hilsman, *To Move a Nation: The Politics of Foreign Policy in the Administration of John F. Kennedy* (Garden City, N.Y.: Doubleday, 1967), 347.

58. John Gittings, *The World and China, 1922–1972* (New York: Harper & Row, 1974), 260–61.

59. Allen S. Whiting, "China's Use of Force, 1950–96, and Taiwan," *International Security* 26, no. 2 (2001): 103–31.

60. Whiting, "China's Use of Force," 111–13.

61. Warren I. Cohen, *America's Response to China: A History of Sino-American Relations* (New York: Columbia University Press, 1990), 191; Gordon H. Chang, "JFK, China and the Bomb," *Journal of American History* 74, no. 4 (1999): 1287–310.

62. Paper delivered in Honolulu, September 20, 1962. United States Department of State, *Diplomatic Papers, 1961–1963*, vol. 22, *China; Korea; Japan.*

63. Norris S. Robert, William M. Arkin, and William Burr, "Where They Were," *Bulletin of the Atomic Scientists* 55, no. 6 (1999): 26–35, www.thebulletin.org/print.php?art_ofn=nd99norris_024.

64. Kochavi, *A Conflict Perpetuated*, 218–20.

65. Noam Kochavi, "Limited Accommodation, Perpetuated Conflict: Kennedy, China, and the Laos Crisis, 1961–1963," *Diplomatic History* 26, no. 1 (2002): 95–135.

66. Arthur A. Schlesinger, *A Thousand Days*, 904.

67. Kochavi, *A Conflict Perpetuated*, chap. 8.

68. Lawrence Freedman, *Kennedy's Wars: Berlin, Cuba, Laos, and Vietnam* (New York: Oxford University Press, 2000), 413.

69. Robert Dallek, *An Unfinished Life: John F. Kennedy, 1917–1963.* (New York: Oxford University Press, 2003); see also: Robert Dallek, "JFK's Second Term," *Atlantic Monthly* (June 2003): 58–66.

70. John M. Murphy, "Crafting the Kennedy Legacy," *Rhetoric & Public Affairs* 3, no. 4 (2002): 577–601.

71. Hebert C. Kelman, "The Social-Psychological Approaches to the Study of International Relations: The Questions of Relevance," in *International Behavior: A Social-Psychological Analysis*, ed. Hebert C. Kelman (New York: Holt, Rinehart and Winston, 1965), 565–607, 582.

72. George Rising, *Clean for Gene: Eugene McCarthy's 1968 Presidential Campaign* (Westport, Conn.: Praeger, 1997), 41.

73. Michael L. Gillette, "Lawrence F. O'Brien Oral History Interview," transcript 6, February 11, 1986, Lyndon Baines Johnson Library, 128.83.78.10/johnson/archives.hom/oralhistory.hom/obrienl/obrien06.pdf.

74. Robert Dallek, *Flawed Giant: Lyndon Johnson and His Times* (New York: Oxford University Press, 1998).

75. F. Charles Parker, *Vietnam: Strategy for a Stalemate* (New York: Paragon Press, 1989), 3.

76. *New York Times*, October 23, 1965.

77. Charles Freeman, Jr., "The Process of Rapprochement: Achievements and Problems," in *Sino-American Normalization and Its Policy Implications*, ed. Gene T. Hsiao and Michael Witunsky (Westport, Conn.: Praeger, 1983), 2.

78. Paige E. Mulhollan, "Dean Rusk Oral History Interview III," transcript, January 2, 1970, Lyndon Baines Johnson Library, 128.83.78.10/Johnson/Archives.Hom/Oral history.Hom/Rusk/Rusk03.Pdf.

79. United States Department of State, *Foreign Relations of the United States: Diplomatic Papers, 1964–1968*, vol. 30, *China* (Washington, D.C.: United States Department of State), www.state.gov/www/about_state/history/vol_xxx/index.html.

80. United States Department of State, *Diplomatic Papers, 1964–1968*, vol. 30, *China*.

81. Paige E. Mulhollan "Dean Rusk Oral History Interview II," transcript, September 26, 1969, Lyndon Baines Johnson Library, 128.83.78.10/Johnson/Archives.Hom/Oralhistory.Hom/Rusk/Rusk02.pdf.

82. T. H. Baker, "Harry McPherson Oral History Interview IV," transcript, March 24, 1969, Lyndon Baines Johnson Library, 128.83.78.10/Johnson/Archives.Hom/Oral history.Hom/Mcpherson/Mcpher05.pdf.

83. T. H. Baker, "Harry McPherson Oral History Interview III," transcript, December 19, 1968, Lyndon Baines Johnson Library, 128.83.78.10/Johnson/Archives.Hom/Oralhistory.Hom/Mcpherson/Mcpher03.pdf.

84. Walt W. Rostow, "Robert S. McNamara Oral History Interview I," transcript, January 8, 1975, Lyndon Baines Johnson Library, 128.83.78.10/Johnson/Archives.Hom/Oralhistory.Hom/Mcnamarar/Mcnamara1.pdf.

85. Mulhollan, "Dean Rusk Oral History Interview II."

86. Harry G. Summers, Jr., *On Strategy: A Critical Analysis of the Vietnam War* (San Francisco: Presidio Press, 1982), 60. See also: Eric M. Bergerud, *The Dynamics of Defeat: The Vietnam War in Hau Nghia Province* (Boulder, Colo.: Westview Press, 1991).

87. Michael L. Gillette, "Ellsworth Bunker Oral History Interview I," transcript, December 9, 1980, Lyndon Baines Johnson Library, 128.83.78.10/Johnson/Archives.Hom/Oralhistory.Hom/Bunker%2De/Bunker1.pdf.

88. Ted Gittinger, "Peter Braestrup Oral History Interview I," accession number 99-03, March 1, 1982, Lyndon Baines Johnson Library, 128.83.78.10/Johnson/Archives .Hom/Oralhistory.Hom/Braestrup/Braestrup.pdf.

89. Dorothy Pierce Mcsweeny, "F. Edward Hebert Oral History Interview I," transcript, July 15, 1969, Lyndon Baines Johnson Library, 128.83.78.10/Johnson/Archives .Hom/Oralhistory.Hom/Hebert/Hebert.pdf.

90. Rene J. Defourneaux, *The Winking Fox* (Indianapolis: Indiana Creative Arts, 1997), 377.

91. Stanley Karnow, *Vietnam: A History* (New York: Penguin, 1983), 15. Note that this was before the Reagan '80s and the reclamation of Vietnam as a "lost crusade."

92. Qiang Zhai, "Cold War International History Project, Virtual Archive. Beijing and the Vietnam Conflict, 1964-1965: New Chinese Evidence," trans. Qiang Zhai, wwics.si.edu/index.cfm?topic_id=1409&fuseaction=Library.Document&Id=244.

93. Mulhollan, "Dean Rusk Oral History Interview II."

94. United States Department of State, *Foreign Relations of the United States: Diplomatic Papers, 1964–1968*, vol. 30, *China* (Washington, D.C.: Government Printing Office), 75–76.

95. United States Department of State, *Diplomatic Papers, 1964-1968*, vol. 30, *China*, 356.

96. Robert A. Diamond, ed., *China and U.S. Foreign Policy* (Washington, D.C.: Congressional Quarterly, 1971), 5.

97. United States Department of State, *Diplomatic Papers, 1964–1968*, vol. 30, *China*.

98. United States Department of State, *Diplomatic Papers, 1964–1968*, vol. 30, *China*.

99. Paige E. Mulhollan, "Walt W. Rostow Oral History Interview I," transcript, March 21, 1969, *Lyndon Baines Johnson Library*, 128.83.78.10/Johnson/Archives.Hom/Oralhistory.Hom/Rostow/Rostow1.pdf.

100. United States Department of State, *Diplomatic Papers, 1964–1968*, vol. 30, *China*.

101. *New York Times*, September 27, 1968.

102. Herbert Y. Schandler, *The Unmaking of a President: Lyndon Johnson and Vietnam* (Princeton, N.J.: Princeton University Press, 1977), 20–22.

103. Museum of the Chinese Revolution, *Album of Paintings* (Beijing: Cultural Relics Publishing House, 1991).

CHAPTER 5

Journey to China, 1968–1973

"I would like to add one point that is very important in terms of our future relations. The Prime Minister has been very forthright in talking about a press story that he saw with regard to our visit to the Ming Tombs. I would only say that I would not reprimand whoever set that up because I liked the little girls. I enjoyed seeing them. But the incident poses a problem that could poison our relations in the future that I think we must avoid. I think that the Secretary of State and Dr. Kissinger and all our party could agree that in the U.S. this new relationship with the People's Republic of China is the big story of the century. We have one thousand newspaper columnists who consider themselves experts. We have one thousand politicians, congressmen, and senators who also will want to comment on this. And they have a right under our system to make statements. They do not consult with us before they make these statements. For example, the story that came out yesterday was that the president had made a decision to recognize Bangladesh at a certain time. We are considering it, but I have not made a decision. The columnist made it up because it was what he wanted. Now there will be stories written by columnists. There will be statements made by politicians that many people abroad will consider to be authoritative and representing the policy of our government."

Nixon to Mao, 1972[1]

1. Nixon: Hedging the Handshake

The year 1968 did not seem a time for great optimism about rapprochement. Chinese and American representatives had met in Geneva and Warsaw 134 times since 1954 with no agreement on any subject since 1955.[2] The record of Richard Nixon—the successful 1968 Republican candidate for president—as an anti-Communist and friend of Chiang Kai-shek was impeccable. He made his early reputation in politics attacking Communism and alleged Communists. In the Senate, he was a member of the China bloc, although not a leader of it. As Eisenhower's vice president he was a coshaper of the iron wall of treaties, troops, and nuclear threats through which Dulles and Ike tried to contain the PRC. We recall the January 18, 1954, *Time* story on Nixon's state visit to Formosa that reported, "Since Nixon's return, and partly as a result of his findings, certain viewpoints are gaining headway in Washington. Among them . . . the U.S. must not even talk about recognizing Red China."

Running for the presidency in 1960 on a 100 percent "no recognition" Republican platform, he tried to paint himself as the true defender against Chinese aggression. On the campaign trail, Nixon told one audience that Kennedy's Jinmen and Mazu position was appeasement in a terrible tradition: "You remember Hitler? We tried to buy him off. First, they gave him Danzig. Then they gave him other parts of Europe. . . . But every time they gave him something he asked for more and more and more, until at last it was war, and the worst war in history."[3] Even after his loss, speaking to the American Society for Newspaper Editors in 1963, Nixon still offered the '50s paradigm of a global Communist leviathan facing the Free World: "Red China and Russia are . . . simply arguing about what kind of shovel they should use to dig the grave of the United States."[4]

Nixon's position evolved in the mid-1960s. He lost a contest for the governorship of California and bitterly announced a retirement from politics. In 1965, he toured Asia; a 1967 trip focused on Southeast Asia. A veteran diplomat recalled that Nixon at that time "pretty well spelled out how we could reach a normal relationship with China, and said that was what we ought to do."[5] Later, in an article for the influential journal *Foreign Affairs*, Nixon claimed that America had "an opportunity to change Communist tactics, to turn it away from external revolutions and inward toward the solution of its domestic problems. . . . [W]e simply cannot afford to leave China forever outside the family of nations, there to nurture its fantasies, cherish its hates, and threaten its neighbors. There is no place on this small planet for a billion of its potentially most able people to live in angry isolation."[6] China's population in 1967 was about 750 million, but the vice president's exaggeration was both rhetorical and tied into an important late-'60s cultural allusion. First, it was an appeal for Americans not to allow China's "masses" to be self-segregated, literally walled off. Second, it was a time of McLuhan's "global village," the launch of the first commercial communication satellites, and the first pictures of earth from space (which actually came from a Russian weather satellite). (Perhaps not by coincidence, the first color images of Earth from space arrived on November 18.) That Earth was a "small" place in a large universe was a pictorial metaphor of wide currency: China's burgeoning population was thus an equally useful symbol for making the case that it could not be allowed to wall itself off.

Running for president in 1968, Nixon was still the conservative, promising "law and order" to a Middle America frightened and angry at the chaos on campuses and in the nation's cities and an interminable war in Vietnam. For the latter, he borrowed from Eisenhower's notebook and hinted of a "secret plan" to end the conflict. He employed General Claire Chennault's Chinese-born widow, a strong Taiwan supporter, as go-between to convince the South Vietnamese government to delay any possible good news about peace talks until after the election. Nevertheless, neither the Republican nor Democratic party platform of 1968 promised any great alterations in China policy. Nixon, however, reaired the need to open China in an interview with *U.S. News & World Report*: "We must not forget China. . . . We must not only seek opportunity to talk with her. . . .We must seek to make changes."[7] Other candidates were making similar statements. As noted, Nelson Rockefeller had advocated "contact and communication" with the PRC.[8] Hubert Humphrey, the Democratic contender, also spoke of "building bridges to the people of mainland China."[9]

Meanwhile, the global situation seemed more unstable. In the summer of 1968, Russian tanks crushed a democratic uprising in Czechoslovakia. In addition, 1968–1969 was a period of increased Sino-Soviet tension. A series of bloody battles took place between border garrisons. The two nations almost came to war, and at this time Russia apparently considered a preemptive nuclear strike against the PRC. It is not clear how much American discouragement of such a move influenced Russia's decision not to go to war.[10]

But Nixon at the time of his election—by a narrow margin—was developing a grand strategy for America in which China played a crucial role. In July 1967, he gave a speech to a conservative group in San Francisco in which he noted, "Ironically, non-Communist Asia, except for Vietnam, is the area which has experienced the most hopeful change. Japan has recovered from the devastation of World War II to the point that its one hundred million people produce as much as Communist China's seven hundred million. Korea, Taiwan, Singapore, Malaysia, and Thailand are all dramatic economic success stories."[11] He was not exaggerating. Given a period of peace, democratic capitalism would eventually win out, but only if America was not overburdened by worldwide commitments. The United States could no longer sustain a global military empire, either in the blood of its soldiers or in its budget. A multipolar world, where a number of nations, not just two superpowers, balanced each other would allow the American economy and polity to survive and flourish. A responsible, engaged China could be one power in the balance for the good in Asia.

A Harvard professor named Henry Kissinger was developing similar ideas. Kissinger had written his doctoral dissertation on Austrian Count Metternich's Congress of Vienna after the Napoleonic Wars, which helped set up a balance of great powers that lasted for fifty years and allowed each nation to maintain relatively small standing armies.[12] (The visual analyst will find revealing the jacket of Kissinger's 1994 summa "Diplomacy," which offers two images: a painting of diplomats at the congress shaking hands and talking, and a color photograph of Kissinger smiling with Mao and Zhou in plush armchairs during a China visit.) In his study of the congress, Kissinger defines one quality of statesmanship as the "ability to recognize the real relationship of forces and then make this knowledge serve his ends."[13] His contemporary concern, like that of Nixon, was in defining the limits of American power and determining how to project them into a stable global balance.

Kissinger noted, presciently, in 1957, "A limited war of any size would absorb so large a part of [American] strategic reserve that we would lose a great deal of flexibility for dealing with any other threat."[14] (This was the case in Korea, would be in Vietnam, and, I argue, is the case in Iraq as of this writing in 2006.) In 1968 he published an essay on "Central Issues of American Foreign Policy" in which he discussed "political multipolarity." At this time the United States and the Soviet Union were the only global military and economic superpowers, but other countries had regional power (such as China toward bordering regions) or disparate points of power (Britain in some of her former colonies). The United States could no longer carry to its maximum strength the burden of containing Communism everywhere. Other nations, growing in prosperity, could be called on to balance one another.

Another key factor influencing the mind-set shift toward rapprochement was that Nixon and Kissinger felt that Communism versus capitalism was no longer the appropriate line of demarcation for friendship and antagonism in the world. As Nixon put it, "Today, the 'isms' have lost their vitality."[15] Or, as Kissinger asserted, "We have no permanent enemies and . . . we will judge other countries, including Communist countries, and especially countries like Communist China, on the basis of their actions and not on the basis of their domestic ideology."[16] For example, America could find a working partner in Socialist Romania but a disagreeable, recalcitrant combatant in democratic France. Kissinger also advocated a revision of American grand strategic doctrine that held that a war with either the Soviet Union or China would automatically precipitate a war with the other.[17] China could be one pole of power, balancing the Soviet Union, but only if it were drawn out from its "angry shell." Later, as

Nixon would put it in a *Time* interview, "I think it will be a safer world and a better world if we have a strong, healthy United States, Europe, Soviet Union, China, each balancing the other, not playing one against the other, an even balance."[18]

At the same time, the elite consensus in America was shifting to support a China initiative. Leading scholars and academics, including MIT's Lucian W. Pye and A. Doak Barnett of Columbia University, wrote a letter to Nixon days after the election calling on the new administration to "move more positively toward the relaxation of tensions between China and the eventual achievement of reconciliation."[19] The League of Women Voters concluded a three-year investigation that recommended resolving the main items of contention with China, such as allowing its admission to the United Nations.[20] (In 1971, the league would focus on Taiwan and conclude that the United States had "only marginal economic and military interest in Taiwan, and should withdraw its troops from there."[21]) In November 1969, a group of congresspeople, including Senate Democratic Leader Mike Mansfield, wrote Nixon a letter calling for a better relationship with the PRC.[22] Senator Henry Jackson, the Democratic hawk, advocated a China opening. In contrast, by 1970, the *New York Times* was able to describe the China Lobby as "moribund . . . a victim of old age and lack of interest."[23]

Public opinion, too, had moderated, even after years of coverage of the horrors of China's Cultural Revolution and China's link to North Vietnam. In 1969, a Gallup survey of January 23–28 found opposition to Chinese admission to the United Nations at an all-time low of 54 percent. In addition, a strong divergence by education level was noted by the survey takers. This had been a minor factor in previous polling but now, for example, 49 percent of college-educated respondents favored the PRC's admission to the UN, while only 21 percent of those with a grade-school education supported the move. The complete sample's support for "going along" if China were admitted to the UN was 56 percent. Yet, ironically, in a September 25–28, 1969, Gallup poll, support for Chinese admission to the UN *dropped* precipitously from the previous year. Only 35 percent of the sample supported such a move, while 49 percent opposed it. China's Vietnam connection probably lay at the root of this tumble, as well as aftershocks of the Cultural Revolution.

Nixon started cautious steps toward a China outreach almost immediately; many were secret, others worded ambiguously or with minimized significance. Nixon and Kissinger shared an understandable mania for secrecy; as we have seen, Sino-American relations had been sabotaged in the past by public diplomacy and press musings. (Both sides were wary of their own extremists attacking the possibility of rapprochement.[24]) In one China-related instruction the president cautioned: "This, of course, should be done privately and should under no circumstances get into the public prints [*sic*] from this direction."[25] Nixon was also preparing for failure. As Denise M. Bostdorff notes, the "surprise" of Nixon's 1971 "I will go to China" announcement was assisted greatly by a "hedging" strategy that "downplay[ed] how quickly rapprochement might come about, and de-emphasiz[ed] the degree of current change in U.S. policy."[26]

But trial balloons filled the skies in 1969, if one cared to look for them. Many would emphasize the "mass" of China's population as reason for better contacts with it. On November 26, 1968, Nixon suggested that the Warsaw talks be restarted; China initially accepted, but then canceled when the United States granted asylum to a defecting Chinese diplomat. Secretary of State-designate William Rogers appeared on January 15, 1969, before the Senate Foreign Relations Committee for his credentials hearing and commented, "It is important to develop channels of communications with Red China, if it is possible and as soon as it is possible, and I think to a large extent that depends on their attitude. Certainly we should be willing to, and I think the president-elect has indicated a very strong desire in that direction."[27] On February 1, Nixon told Kissinger, then an adviser to the National Security Council, to "give every encouragement to the attitude that the administration was exploring possibilities of rapprochement with the Chinese."[28] Nixon also made some further minor advances in liberalizing trade and travel to China.

Other developments seemed to favor the enterprise. In March 1969, two major border battles took place between China and the Soviet Union. (Between 1960 and 1970 the Soviet Union had more than tripled its military strength along the Chinese border.[29]) Nixon and Kissinger felt the conflict would be an incentive for the PRC to seek contact with the United States.[30] On April 21, Secretary of State Rogers addressed the annual luncheon of the Associated Press in New York City and stated, "One cannot speak of a future Pacific community without reference to China . . . .Not even a nation as large as mainland China can live forever in isolation from a world of interdependent states. Meanwhile, we shall take initiatives to reestablish more normal relations with Communist China and we shall remain responsive to any indications of less hostile attitudes from their side."[31] Nixon also passed on a conciliatory message to the Chinese leadership via French president Charles de Gaulle, who visited the PRC in April.[32]

On May 15, the National Security Council's Senior Review Group, chaired by Henry Kissinger, reviewed China policy. The summary of the meeting shows Kissinger stating, "Everyone agrees that we wish to reduce the risk of war with seven hundred million people, but the question is whether alignment with the Soviets, more conciliatory posture toward China, or some combination would best achieve this end."[33] In the summer of 1969, the Soviet Union published dark hints about a nuclear attack on China and also quietly explored the option with the United States.[34] In mid-August 1969, the Soviets made another "testing the waters" proposal of destroying China's nuclear facilities jointly with the United States; Nixon and Kissinger rejected the idea as out of hand.[35] On October 29, Kissinger sent a memorandum on foreign policy to the president. He argued for "the gradual improvement of relations with Communist China while continuing to maintain the strength of American deterrence against direct Chinese military aggression and nuclear blackmail."[36]

Earlier in September, Kissinger, with Nixon's approval, told America's ambassador to Poland to contact his Chinese counterpart and "tell him that we were prepared for serious talks."[37] The contact actually took place at a Yugoslav fashion show held in Warsaw on December 3, but the Chinese ambassador would give no answer, pleading that he had no instructions from his home government. (Zhou would joke later that the surprise offer almost gave their ambassador a heart attack.) Almost a week later, however, the American ambassador was invited to visit the Chinese embassy. This became the first formal contact between the nations inside an embassy since the fall of China to Communism in 1949. After some stops and starts, the Warsaw talks resumed in July 1970.

Changes in China brought about this turnaround. Edgar Snow noted that the PRC in 1969 was exuding less heated rhetoric about America than it had for some time previously. In an article in the *New Republic*, Snow declared that "China's door may be open a crack or two now."[38] The Soviet conflict did not immediately throw China into America's arms, but it was an important motivator to consider rapprochement with the old enemy, America.[39] Before the first border clashes, Mao had sensed his nation's vulnerability if it found itself at war with two superpowers. He commented to a Western visitor, "Our country, in a sense, is still a non-nuclear power. With this little nuclear weaponry, we cannot be counted as a nuclear country. If we are to fight a war, we must use conventional weapons."[40]

In the summer and fall of 1969, Mao consulted with a number of top political and military leaders in the country, asking for a full policy review. Among their advice was that, if China faced the possibility of either war or a sustained trouble with the Soviet Union, then the "card of the United States" should be used, perhaps via public negotiations.[41] At this point, Mao's perennial paranoia may have actually assisted the cause of rapprochement. Surrounded by enemies, this time on his northern border as well, Mao said to his physician in August 1969:

> Think about this, we have the Soviet Union to the north and the west, India to the south, and Japan to the east. If all our enemies were to unite, attacking us from the north, south, east, and west, what do you think we should do? . . . Think again, beyond Japan is the United States. Didn't our ancestors counsel negotiating with faraway countries while fighting with those that are near?[42]

Mao developed this musing into a theory of principal versus secondary enemies. The former were the enemies that threatened the lifeblood of the nation, and all seriousness should be attached to defending against them. Secondary enemies, such as the United States, could be dealt with more openly, perhaps even coming to some sort of accommodation. Certainly, when Mao justified Nixon's visit to his own people, he did so by saying that the rapprochement was "another tactic in the struggle against imperialism."[43] America's sagging fortunes in Vietnam had also convinced the Chinese leadership that by 1969 the United States was, according to Kissinger, the "lesser threat" compared to the USSR[44] Indeed, China could see that American involvement in the Vietnam War was winding down with Nixon's "Vietnamization" strategy.[45] China began to see itself as more in conflict with the Soviet Union, which it actually bordered, than the more abstract enemy, the United States. Zhou commented later, "That we oppose two superpowers is a slogan. The essence lies primarily in opposition against the most actual enemy, Soviet revisionist social-imperialism, and in the combat against this social imperialism."[46]

For the second time in this era, Mao tried to use Edgar Snow as a symbol of his wish to ameliorate relations with America. On the October 1, 1969, festivities of the anniversary of the founding of the People's Republic, Mao took the unprecedented step of inviting Snow to pose next to him on the reviewing stand at Tiananmen.[47] In the *People's Daily* photograph that appeared the next day, the emphasis was even more striking, since Zhou Enlai had ordered the newspaper's editors to crop out everyone else from the image except Mao, Snow and his wife, and an interpreter. Again, the United States missed the symbolism—a curious failure, considering that Kremlinologists and Sinologists had developed a subscience of analyzing and interpreting the various presences, absences, and positions of Communist leaders on reviewing stands or at funerals.[48] Kissinger, however, wryly noted in his memoirs that the Chinese had "overestimated our subtlety, for what they conceived was so oblique that our crude Occidental minds completely missed the point."[49] In another interview with Snow on December 18, Mao praised Nixon's honesty and forthrightness, and even directly invited Nixon to visit without precondition or with a demand that any issues be settled.[50]

Meanwhile, on July 9, 1970, Secretary of State Rogers made a statement on Japanese television that China was the most important nation in deciding the future of Indochina. The Chinese responded quickly by restarting the Warsaw ambassadorial talks.[51] Kissinger shortly afterward secretly met with the PRC ambassador to France in Paris. Negotiations proceeded rapidly toward establishing a principle of "curbing Moscow's geopolitical ambitions."[52] President Nixon met with Romanian president Nicolae Ceauşescu on October 26, 1970, during the latter's state visit. Romania at that time was going down a quasi "Tito road" of independent foreign policy from Moscow, and its leader made a link between East and West. Ceauşescu and Nixon discussed the China situation. The State Department Report summary indicates that Ceauşescu reported "that China desires to have improved relations with the United States and is ready at any moment to occupy its place in the UN, including this year." Nixon later in the meeting remarked, "American policy is one of wanting friendly relations with both the U.S.S.R. and eventually with Communist China."[53]

1971 was, of course, the fateful year of the door opening, and again the arguments in favor of rapprochement tended to appeal to a vision of the China as a large number of people who needed to be deisolated. Nixon gave a State of the World report on February 25, 1971, and declared, "The twenty-two-year-old hostility between ourselves and the People's Republic of China is another unresolved problem, serious indeed in view of the fact that it determines our relationship with 750 million talented and energetic people."[54] When the first Warsaw talk of his presidency took place on March 7, 1971, the American proposals actually included some items that the Chinese had floated during the Eisenhower administration, only to be shot down by Dulles, such as an exchange of journalists. Wang, still the Chinese representative at the talks, stonewalled any agreement. For China, he reiterated, there could be no agreement on any item until the question of Taiwan was resolved.[55]

Other developments contributed to more subtle but important contacts. On March 15, 1971, the United States lifted all remaining travel restrictions to the PRC. At the same time, the Chinese national table tennis team traveled to Japan to play in the thirty-first World Table Tennis Championship. Mao had personally intervened to make sure that they attended, although he was worried about their safety (it is not clear from what danger). The relationship between the Chinese and American players became a friendly one, and within a month the PRC extended an invitation to the American team to play exhibition matches in the PRC.[56] Kissinger understood the "ping-pong diplomacy" to hold a deeper meaning and responded with a relaxation of trade restrictions.[57] Seven Western journalists were granted visas to cover the athletes' trip. Zhou hosted the group in royal honor in the Great Hall of the People and told them, "Your visit at our invitation has opened the door to friendship between the peoples of the two countries."[58]

The real breakthrough came when the Pakistani president passed on, on June 2, 1971, the following information, which would prove to be the first communication that would radically change Sino-American relations:

> Premier Zhou Enlai has seriously studied President Nixon's messages . . . and has reported with much pleasure to Chairman Mao Zedong that President Nixon is prepared to accept his suggestion to visit

> Beijing for direct conversations with the leaders of the PRC. Chairman Mao Zedong has indicated that he welcomes President Nixon's visit and looks forward to that occasion when he may have direct conversations with His Excellency the President, in which each side would be free to raise the principal issue of concern to it. . . . Premier Zhou Enlai welcomes Dr. Kissinger to China as the U.S. representative who will come in advance for a preliminary secret meeting with high level Chinese officials to prepare and make necessary arrangements for President Nixon's visit to Beijing.[59]

Kissinger's secret visit to China in early July worked out a basic agreement and a communiqué that removed preconditions to further talks. Anything that might hurt the relationship was suppressed. For example, the USIA had just completed a film on the Chinese occupation of Tibet; Nixon ordered the film shut away. (In 1972, all military and financial assistance to Tibetan resisters was stopped.) On July 15, 1971, Nixon spoke live to a national audience, revealed the contacts, and announced his upcoming journey to China, concluding:

> I have taken this action because of my profound conviction that all nations will gain from a reduction of tensions and a better relationship between the United States and the People's Republic of China. It is in this spirit that I will undertake what I deeply hope will become a journey for peace, peace not just for our generation but for future generations on this earth we share together.[60]

Kissinger queried rhetorically, "The announcement that shook the world? It was a revolution in diplomacy—a geopolitical revolution."[61] Indeed, the backdoor and back channel took everyone from China scholars to news editors by surprise. Nixon, of course, kept Congress uninformed about his overtures; even after the announcement he simply told congressional leaders that they should let him work unimpeded. And members of his own government, like Secretary of State Rogers, were mocked for not being "in the loop."

The next day, Kissinger briefed members of the press on his trip to China and the upcoming presidential visit. His main goal was to downplay any expectation of a major breakthrough in relations. He, too, however, appealed to the "mass" of China:

> Let me make one other point: nothing that has been done in our relations with the People's Republic of China has any purpose or is in any way directed against any other countries, and especially not against the Soviet Union. We are taking these steps because we cannot imagine a stable, international peace in which a country of 750 million people is kept in isolation. We believe that by improving re-

> lations with the People's Republic of China we are contributing to peace in the world, and therefore are contributing to all nations.[62]

On July 19, 1971, Kissinger met with Soviet ambassador Anatoly Dobrynin and found him "for the first time in my experience, totally insecure." The Soviet diplomat was concerned about China's relation with his country and even stated, according to Kissinger's notes,

> What was there really to talk about between us and the Chinese? Were we interested in Chinese domination of Southeast Asia? He had always thought that the Soviet interests and ours were much more nearly complementary with respect to the defense of Southeast Asia. I said that I wasn't certain that the Chinese had aggressive tendencies in Southeast Asia but that in any event we would not favor Chinese expansion beyond their borders.[63]

As Kissinger relates in his memoirs, the payoff in Soviet relations by the China move was tangible and quick: "Other negotiations deadlocked for months began magically to unfreeze . . . within weeks of the Beijing announcement."[64]

On October 20, 1971, Kissinger flew again to China to work out the agenda and protocol for Nixon's trip. At the same time, the United States supported a UN vote to seat China. However, in spite of U.S. opposition, Taiwan lost its seat in the United Nations by a lopsided vote on October 25. There was much guessing among journalists on the effect of Kissinger's second trip to Beijing on the UN vote. For example, Seymour M. Hersh writes, "Photographs of his sight-seeing in Beijing made page one in New York and obviously affected the U.N. delegates, who voted on October 25 to seat the PRC." Hersh goes on to say that China's satisfaction at its adroit handling of the Nixon administration was made clear in September 1971 by Huang Hua, who had aided Zhou Enlai in his negotiations with Kissinger. "The Americans say they are going to step down from some of their high responsibilities in Asia," Huang told the journalist Wilfred Burchett. "We are happy to provide a small stool for them."[65] The floodgates of China opening up were now a global phenomenon. The Canton Trade Fair of 1971, for instance, drew ten thousand businessmen from fifty nations.[66]

The effects on American public opinion were mixed and must be interpolated with attitudes held toward Nixon and his policies on other issues such as inflation, a reelection victory, and then Watergate. Mid-1971 was a low point of popularity for the president, as opposition to Vietnam from both left and right climbed and bad economic news unfolded. As said, 1969–1970 saw drops in the estimation of the PRC by Americans. But in a Gallup survey of May 14–17, support for Communist China's admission to the UN stood at 45 percent, with 28 percent opposed. In a Gallup poll of October 19–24, 1971, Communist China was still named as "a greater threat to world peace" than Russia by 56 percent of the sample. In other polling, support for Nixon's projected 1972 Democratic opponent Edmund Muskie of Maine began to drop relative to that of the president.[67] In a Roper survey conducted October 18–27, Americans were specifically asked about Nixon's China engagement. 51 percent of the 1,499 respondents said they approved of his "decision to visit Communist China"; only 22 percent disapproved. On the other hand, 36 percent approved of "his decision to support Communist China's entry into the United Nations," while another 36 percent disapproved and 28 percent expressed no opinion. 17 percent of the sample, when asked which countries "are most important for us to have as military allies," cited Communist China. Only 9 percent listed Communist China when asked, "Which do you think are most important to us in terms of trade?" However, China was listed by 51 percent of the sample as the "greatest threat to world peace." 0 percent of the sample named China when asked, "If for some reason you couldn't live in the United States, which [country] would you be most inclined to move to?"

Americans were not willing to pay the price of better relations with Communist China by abandoning Taiwan. When posed with the dilemma of admitting Communist China to the UN and expelling Nationalist China versus continuing to keep out Communist China in order to retain Nationalist China's membership, 62 percent of the sample were willing to keep the PRC out in the cold. Only 8 percent were willing to expel the Nationalists in order to allow in the Communists. In short, there was no radical "pro-PRC" shift in public opinion. Perhaps "wait and see" might be the best description of the nation's mood, with "seeing" a new vision of China as the crucial factor.

The planning for the China trip on both sides was extraordinary although, to the initial chagrin of both Kissinger and the Chinese, Nixon's advisers quickly concocted elaborate proposals to make the trip, as one aide put it, into "one of most exciting events in TV history."[68] As print journalists would later complain, the emphasis of the White House was a visual spectacle for television, and cooperation with the major TV networks was close.[69] Moreover, this was *Nixon's* trip to China: all focus would be on the commander in chief as superstar.

Nixon arrived in Beijing on February 21, 1972, the first American president to visit China. His glee was obvious as he descended from the *Spirit of '76* presidential jet with his wife, Pat, to shake the hand of Zhou Enlai.[70] (Nixon had ordered his entourage to stay inside the plane so that the television spotlight would be solely on the pair.[71]) In turn, Zhou got the photo op of his choice: against diplomatic precedent he waited stiffly for Nixon to proffer a hand at the bottom of the ramp, thus visually atoning for the famous nonhandshake of the Dulles years. The two then exchanged pleasantries, each visibly triumphant at getting the moment he wanted.

So began, as Nixon put it, "the week that changed the world."[72] The president and his wife visited what are now the standard tourist targets for anyone seeing Beijing—the Ming Tombs, the Great Wall, Tiananmen Square, and the museums. For most Americans in 1972, however, it was all new, as if the Nixons were touring Wonderland and taking the nation along. Even a nine-year-old future China lover (me) was fascinated by the television coverage of Nixon's visit. I recall my father, a professor of international management who had worked on Asian affairs, deeming it "the world's most expensive travel show." Later, cynics would say the touristy aspects of the meeting were to foreground show over substance, but in fact the spectacle was a necessary first step to clear the air of twenty-two years of antagonism.

The capstone event came when Nixon met Mao in his residence within the Great Hall of the People. The handshake with the chairman was obviously such an attractive moment to both men that it was one of the longest diplomatic handshakes in history, lasting about a minute.[73] The drama of the background to the event from the Chinese side has only recently been revealed. Mao's private physician, Li Zhisui, provided the forensic detail which had been an object of speculation by American journalists: "In the picture of Mao shaking hands with Nixon that was published, both were smiling broadly, and the Chinese report described Mao as energetic and glowing with health, his face flush with color. Many noted that Mao had gained weight and took that as a sign of good health, too. The American press, knowing that Mao had been ill and was having trouble speaking, speculated that the chairman had had a stroke. But the Chinese and the Americans were both wrong. Mao had not gained weight. He was bloated by edema. He had been suffering from congestive heart failure, not a stroke."[74]

Mao had gone through a grueling health regimen in order to look presentable to the American president and to the world's press. Nevertheless, Li notes, "He was still so bloated that he had to be fitted with a new suit and shoes. His throat was still swollen. And he had difficulty talking. His muscles had atrophied from weeks of immobility, so we put him on an exercise routine a week before President Nixon's arrival. He practiced sitting down and getting up, and an attendant guided him slowly around the room to get him used to walking again. Mao was as excited as I had ever seen him the day Nixon arrived."[75] In case Mao collapsed, medical equipment such as oxygen tanks and emergency personnel were hidden discreetly near the meeting rooms. There was a brief comedy, according to Li, when Nixon entered for his meeting with Mao: the doctor greeted him and showed him to Mao's study, but then stepped away to check on the status of the medical preparations. As a result, Nixon's entourage temporarily lost the president, and the tin roof of the room blocked communications. One of the Chinese interpreters assured the American Secret Service that Nixon was safe with Mao.

Li remarks that one point of the conversation revealed Mao's pragmatic streak. Earlier, he had spoken well of his archnemesis John Foster Dulles because of his consistency and toughness, and disparaged the "mixed signals" of the Truman administration. Now, it was clear that Mao had a similar respect for Nixon's record as an anti-Communist and his resolution as a statesman: "He speaks forthrightly—no beating around the bush, not like the leftists, who say one thing and mean another."[76] Mao, also to Nixon, expressed praise for American hard-liners, admiring their qualities of toughness and resilience.[77]

The negotiation sessions were cordial; the notes of the meeting read like two groups of people genially trying to move beyond prior ill will with not a few apologies, excuses, and self-blaming.[78] In one session with Zhou at the Beijing airport on Saturday, February 26, 1972, Nixon stated, "I think it can also be said that we do have differences, and you can't build a bridge covering sixteen thousand miles over twenty-two years in one week." Zhou, for his part, noted one misunderstanding with communication and contacts on the trip. American reporters had complained that the Chinese adults and children at the Ming Tombs were stage props, sent along so that Nixon might have a photoop interacting with them. Zhou admitted:

> It was quite unnecessary to put up a show in the Ming Tombs. . . . Some people got some young children there to prettify the Tombs, and it was putting up a false appearance. Your press correspondents have pointed this out to us, and we admit that this was wrong. We do not want to cover up the mistake on this, of course, and we have criticized those who have done this.[79]

Nixon graciously—and anyone who thinks Richard Milhous Nixon was incapable of grace (or graciousness) should read these transcripts—went on to gently explain to the Chinese what he felt would be a problem in future relations. Again, Nixon believed that China was "isolated." Kissinger had noted that China was a power without much knowledge of the workings of other kinds of sociopolitical systems. Nixon also emphasized to Zhou that the American press would go its own way, and what the Chinese leaders read in it should not be taken as directed from the White House. In fact, he emphasized that Chinese leaders should not regard American newsmen as heralds of his policies.

The major public outcome of the trip was the Shanghai communiqué, which listed some policy differences but also enumerated common ties, and put aside the Taiwan issue.[80] The key point on Taiwan was the statement: "The United States acknowledges that all Chinese on either side of the Taiwan Strait maintain there is but one China and that Taiwan is a part of China. The United States government does not challenge that position. It reaffirms its interest in a peaceful settlement of the Taiwan question by the Chinese themselves. With this prospect in mind, it affirms the ultimate objective of the withdrawal of all U.S. forces and military installations from Taiwan. In the meantime, it will progressively reduce its forces and military installations on Taiwan as the tension in the area diminishes."[81]

For several decades, this was the historical "transcript" of Nixon's trip to China: a breakthrough of a symbolic kind, a masterpiece of public relations for both the Chinese and the Americans. But in recent years, National Security archives and Chinese sources have detailed that there was *real* substance in the many meetings. An entire book could be written—and no doubt will be—on this particular topic. But to give two examples of substantive dialogue, (a) the United States provided the Chinese military with the deployments and breakdown data of every Soviet military unit at or near the Chinese border, and (b) Henry Kissinger assured the Chinese that the United States had no intention or wish to overthrow the Communist government of North Vietnam and, further, all but admitted that the United States had few if any interests left in preventing the fall of South Vietnam to Communism. That these were golden gifts to the Chinese government is an understatement, and only the passage of years has made them less startling in the recapitulation. The key point, from the perspective of this study, is that if any of the "leaks" of information and opinion to Mao's government by Nixon's government had become public knowledge then—that is, reported in the press—it would have caused a major diplomatic disaster in Soviet-American relations and the possible fall of the Nixon presidency after a domestic revolt among practically almost all his supporters. That such a "Chinagate" scandal never occurred, thanks to Kissinger's mastery of secret diplomacy (at least in this case), gives further credence to the fact that so little of what we were hearing, seeing, and reading about China during the heady days of '72 and '73 was based on true insider knowledge of what was really going on.

In the wake of the trip, the scope of American contacts and relations with China expanded. In 1972, the PRC bought nineteen million bushels of American wheat,[82] but made bigger news when it bought ten Boeing 707 transports and $150 million worth of spare parts.[83] Overall trade in 1971 was a paltry $5 million a year. This figure jumped to $95.9 million in 1972, over $800 million in 1973, and nearly $934 million in 1974.[84] A U.S. liaison office was set up in Beijing, and although it was not yet called an embassy, it performed all the same functions. The signing of a peace treaty in Vietnam and the evacuation of American troops from Indochina reduced the second flash point of U.S.-China relations, fundamentally leaving only Taiwan as an issue.

The image of China shifted radically as well. The Nixon trip was a photo op, but an enormously successful one. In the initial adulatory writings of many prominent American journalists, it was as if China had metamorphosed magically the moment Nixon's plane set down.[85] There continued to be an opposition to recognition, but not an organized one. Stanley D. Bachrack, the author of the definitive study of the leading China Lobby group, notes that "the death knell for the Committee of One Million came on the night of October 25, 1971, in the great hall of the General Assembly of the United Nations" when Nixon shook Mao's hand.[86] Many conservatives were caught up in the fervor. Joseph Alsop, a great friend of the Chiangs, was one of the biggest boosters of a reformed vision of the People's Republic. He wrote, "Everything in China has changed, in truth, except the endlessly resilient, hard-working, and clever Chinese people. The quality of life has changed, vastly for the ancient ruling class but for the better for everyone else."[87] A "China fever" took hold in the United States, marking a return to a previous era (the 1700s) of fascination and idealization.[88] As detailed below, there was a minor explosion of China fashions and Chinese goods in America. Chinoiserie, the mania of China collecting, had finally made its comeback.[89]

In terms of public opinion, in 1972, in a survey conducted March 3–5, 18 percent of a Gallup sample felt that "President Nixon's trip to China will be 'very effective' in terms of improving world peace." 50 percent felt that the trip would be "fairly effective" toward that goal, while 24 percent felt it would be "not at all effective." 98 percent of the sample had heard about Nixon's trip to China. This large group was then asked, "What impressed you most about the trip?" The responses given most often were: "(1) the opportunity for improved relations between the two countries—and for world peace; (2) the warm reception given Nixon and U.S. officials; (3) the lifestyle of the Chinese—their 'serenity, obedience, ability to work hard, primitiveness.'" Fully 74 percent of the sample felt that "hard-working" was the favorable term that most described the Chinese people. In a 1973 Roper poll conducted September 28–October 6, 53 percent of the sample approved of Nixon's "handling of relations with China."

China seemed more in the news than ever in 1972 and 1973. Almost as if by design, a new iconic animal was offered to America to substitute for the dragon so often employed as a China metaphor in *Time* issues of the past: Zhou Enlai sent two giant pandas, the first of many, as a goodwill present for "the American people." (America returned the favor with two musk oxen; these animals were among the only trade goods that Qing Dynasty China had sought from young America.) On May 22, Nixon traveled to Moscow for a summit with Soviet leaders, who continued to be as flexible as American negotiators had ever recalled them. On May 26, President Nixon and Leonid Brezhnev signed the first Strategic Arms Limitation Treaty (SALT I); détente had begun, and some would say that the other long road of the Cold War, leading to Gorbachev, perestroika, and ultimately the collapse of the Soviet Union, was started as well.*

* On June 17, there was one minor domestic news item that would eventually influence China policy: some men were arrested trying to break into the Democratic Party headquarters at the Watergate apartments in the nation's capital.

Days later Kissinger again flew to Beijing for a seven-day visit "to promote normalization." A week later a party from the U.S. House of Representatives arrived for their own China visit; it was not well covered in the press. A week before the presidential election, the White House, citing an imminent "peace" in Indochina, stopped the bombing of North Vietnam. Nixon won reelection in a landslide. China, notably, was not an issue in the contest between the president and George McGovern. Later in the month, the United States lifted all further travel restrictions to China. Then, as Vietnam peace talks stalled, at the end of December the United States began a massive "Christmas bombing" of North Vietnam. Although the action provoked an outcry at home, the result was a peace treaty within weeks, one which left South Vietnam penetrated by Northern units and actually just providing a fig leaf for full American withdrawal.

In 1973, Kissinger traveled to China several more times, and the nations established a normal postal service. In a November trip Mao admitted that, like the American president, he was consumed by internal issues.[90] Mao was at his most conciliatory and told Kissinger, "I say we can do without Taiwan for the time being, and let it come after one hundred years."[91] Events in China, however, continued to undermine an immediate friendship between that nation and the United States. Reports about the "internal issues" suggested that the hard-left radicals were putting a lot of pressure on the moderate wing, of which Zhou was identified as the leader. Nevertheless, in February, China and America agreed to open official liaison offices in their respective capitals—there were functioning embassies in all but title—and appoint diplomats of ambassadorial rank to head them. By May the offices were up and running. This was a significant step for the PRC, in that it had originally requested immediate recognition by an actual embassy and fully expected the delay toward full normalization of relations to be a brief one.[92]

The wild enthusiasm that greeted the Nixon visit to China was followed by the traditional corrective slump in mood after China did not transform overnight.[93] Reporters, feeling self-conscious at overindulging in the praise of China, wrote columns about other reporters overindulging in praise of China. Many in Congress and elsewhere chafed at Nixon's initial secrecy, although few could argue that the trip could have been planned in the first place without it. But Nixon's reputation for conservatism helped him domestically in two ways from losing any support due to the China move. First, it made him less suspect as a radical by moderates and even some liberals. Second, with the leftward tilt of the Democratic party, as Stephen E. Ambrose described, the hard right "could rant and rave, but they had nowhere to go in a national election except to Nixon, which would force them to mute their criticisms."[94] Nixon would ultimately resign from office, but not due to his China initiative. Going to Beijing would always be cited as the great achievement of his presidency, even though it would be some years before relations between the two countries were truly normalized.[95]

2. Pictures: The Long Visit

The overall spread of China pictures by year from 1968 to 1973 in *Time* is shown in table 5.1. Unsurprisingly, China news pictures follow the curve of China news. Nineteen seventy-one and 1972 are among the richest years for seeing China in the publication both in raw numbers and in percentages of the total dataset of pictures from 1949 to 1973. As shown in table 5.2, China also receives among the highest numbers of cover stories in *Time's* history in 1971 and 1972, and as recorded in table 5.3, the total surface area of China imagery is also peaking in those years of the Nixon announcement and the trip. But the most radical insight the overall numbers provide is the ideological "view" of China in the years from the end of the straits crisis through the beginning of the post-Nixon trip. In table 5.4, we see the rate of "social dysfunction," that is, the negative views of China, *peaking* in 1970, then collapsing in the subsequent three years to the lowest numbers so far in the sample. Indeed, 1972 is the year China received the least negative visual coverage by *Time* magazine *for the entire run of the analysis.* Again, 1969–1970 were not years when China's image was getting better, although elite voices may have been calling for better contacts. Nixon led, and did not follow, the nation, and the press radically reversed course in China coverage, even during a "backlash" against the adulation of the Nixon voyage.

China is almost completely absent in *Time's* coverage of 1968 except as a periphery in maps of Indochina. The Tet Offensive, the crushing of the Prague Spring in Czechoslovakia, and the American presidential campaign eclipsed other news. There are nevertheless a few caustic pictures of China: A cartoon from a Dutch newspaper in the May 10 issue accompanies a story that details all the various infighting among Communists, including the beginning of the Czechoslovak uprising. The image shows various Communist personages, including Mao, punching, kicking, scratching, and biting one another while a bearded leviathan Karl Marx looks upon them perplexedly. The connection is the 150th anniversary of Marx's birth. The accompanying story comments, "Marx

Table 5.1. China Pictures by Year, 1968–1973

Year	*Freq.*	*Percent*
1968	9	0.5
1969	32	1.7
1970	22	1.2
1971	87	4.7
1972	98	5.3
1973	29	1.6

Table 5.2. *Time* China Covers by Year, 1968–1973

Year	*Covers*
1968	0
1969	0
1970	0
1971	3
1972	4
1973	0

foresaw a classless society under Communism and predicted that the State would wither away; but Communism seems to have produced just the opposite. Russia, China, and most of the other Communist nations are run by a tiny elite that holds stronger sway over workers than Marxists' capitalist villains ever did." *Time* devotes some space to chaos in China and the collapse of a Communist insurgency in Malaysia and the seizure of the Pueblo by China's ally, North Korea. Even the Chinese opera is held up for ridicule, with two images of earnest performers accompanying a story that depicts the musical form as "insipid water torture." One film made from an opera is described to be "as exciting as a Communist indoctrination lecture, which is what it is."

One item in the May 3 issue, however, illustrated by a placid scene of an old city surrounded by mountains, gives a slight hint of a view of China that will impress a number of Western visitors a few years in the future. A story reports a publication by a French team of a modern-day travel guide to the People's Republic. Among the "meaningful information about the vast xenophobic country" that appears in the book is, "The visitor should not seek to discard anything of even residual value in China. It is bound to pursue him until the Chinese can

Table 5.3. Surface Area of China Pictures by Year, 1968–1973*

Year	*Freq.*	*Percent*
1968	159	0.3
1969	616	1.4
1970	679	1.5
1971	2,125	4.7
1972	2,240	4.9
1973	553	1.2

*This table shows the yearly surface area of China pictures in *Time*. The calculation of surface area is based on:
Two pages = 187 square inches.
One page = 93.5 square inches.
Half page = 46.75 square inches.
Quarter page = 23.38 (rounded) square inches.
One-ninth page = 10.39 (rounded) square inches.

Table 5.4. Social Dysfunction in the PRC by Year, 1968–1973*

Year	*Percent*
1968	44.4
1969	56.2
1970	77.3
1971	29.9
1972	11.2
1973	13.8

*This table shows the percentage of pictures of social dysfunction in the PRC in *Time* by year in relation to the total number of pictures of the PRC.

return it." This minor notation will become one of the major praises of China by journalists and the first visitors to the PRC with Richard Nixon and thereafter. It is an important footnote because it is also what visitors to China in the '70s recall most vividly. One journalist recollected abandoning a pair of worn-out shoes in a Shanghai hotel and having them continually returned to him by the staff every place from which he tried to discard them anew. Such a verbal picture contrasts with the chaos that most Chinese imagery of 1968 and 1969 portrays. We can see one of the reasons why Nixon's China trip unleashed a flood of praise for the PRC even though the nation was, objectively, no less under the thumb of an evil tyranny in 1972 than it was in 1968. The China that we would see in pictures, and the one that the chief shapers of China's image would encounter, in restaurants, hotels, and on the street, was of an ordered, honest, neat people; these were not the fetid street scenes of Republican Beijing and Shanghai, nor the Red Guard mobs and starved peasants. That both those realities had been swept aside purposely to present China's best face to the world almost didn't matter.

Time's coverage of China in 1969, the first year of Nixon's presidency, was not a remarkable departure from its previous portrayals. The telescope of retrospective history, however, allows us to perceive some changes on the horizon. The first China image is associated with a story that does not seem to bode well for improvement in Sino-American relations. The picture appears in the February 14 issue; it features a photo of a smiling Asian man who is identified as a defector from Communist China. In the accompanying story, *Time* notes that the defection "is expected to set off another round of witch-hunting in China's foreign service. . . . Beijing assailed Richard Nixon as a 'jackal' and demanded that" their renegade diplomat be returned. As noted above, we know that this incident did indeed set back resumption of the ambassadorial talks in Warsaw. More bad news from China is reported in the February and March issues on Sino-Soviet border skirmishes. Interestingly, the articles make no speculation that China might attempt to play an "America card" against the Soviets, or that the Sino-Soviet split is a diplomatic opportunity for the United States.

These speculations are also absent when, on March 21, *Time* devotes a story to "Moscow v. Beijing: Offensive Diplomacy." The accompanying photos, subcaptioned "No longer socialist brothers but instead dangerous foes," show "Soviet frontier guards view[ing] comrades' bodies" and a blurry in-your-face confrontation between border guards. A cartoon taken from a French newspaper shows the head of Karl Marx with its two eyeballs replaced by Mao and Brezhnev glaring across at each other. Here the speculation in relation to U.S. diplomacy again fails entirely to predict what would actually come to pass: "The time may come when the U.S., either as a counter-balance to Russia or to aid a more moderate and rational post-Mao China back into the community of nations, will have to assuage China's fears." In short, *Time* cannot project Mao's opening China.

Stories in the weeks to follow detail events in China such as meetings of the CCP leadership and the Sino-Soviet split's effect on other Communist parties throughout the world. *Time*'s June 20 issue devotes a photo-essay to "Where Russia and China Collide: Build-Up on the Soviet Border." These are unsensational pictures of a few military installations and Russian soldiers standing around, with China as the far bank in the background. As the weeks pass, there is further speculation about the Sino-Soviet split and its effect on Indochina policy, especially for North Vietnam, culminating in the September 19 report on the funeral of Ho Chi Minh.

"Two Decades of Communism" in China are given a curiously small space in *Time*'s October 3 issue. This is perhaps one of the clearest signs of the passing of Henry Luce. One could imagine that with Luce as editor in chief, *Time* would have devoted a cover to an enumeration of Communist failures. The story is restrained; many of the disasters that afflicted China are described and are blamed on the leadership, but the tone is calm. A picture of "Red Guards Hailing Mao Zedong" dated 1966 is subcaptioned not, as we might expect from the Luce era, as some kind of agents of horror, but "prescriptions of a revolutionary romantic." This is perhaps the most benevolent description of Mao offered by *Time* in twenty years. The story concludes with a long quote from Edwin O. Reischauer, then a professor at Harvard and a former U.S. ambassador to Tokyo. Within it—and since it ends the story without rebuttal, we may surmise that it embodies *Time*'s musings about the future—the Chinese "mass" returns, not in anger but in sadness, and with a resurrection, at least implicitly, of the old American hope of China's changing for the better, in the Western definition of improvement. "In the long run," Reischauer writes, "the chief problem that China presents may not be the danger that it will be so rich and strong, as well as hostile, that it menaces our basic interests, but rather that it may fall so short of meeting the economic needs and aspirations of its people that it remains an unstable and sick fifth of humanity." *Time* adds, "Not until Beijing's leaders begin to busy themselves with the task of satisfying those basic needs will China be able to set out on the long road that Mao talked about twenty years ago."

In the concluding pictures and descriptions of China in *Time* in 1969, a slight shift occurs which, in retrospect, provides a hint of one possible future that did come to pass. The Sino-Soviet split is depicted as easing slightly from the brink, with rhetoric and border provocations pulled back. A story from the October 10 issue concludes that China was a "puzzle," and that "the most striking thing about Communist China as it begins its third decade [is that] though it is the world's most populous nation, it has drawn so tight a curtain around itself that virtually nothing of its present policies, personnel and problems is known for certain." This is a remarkable statement for *Time*, which in the past had claimed that it was very difficult to uncover what was going on in China but assured us that its diligence and host of experts would proceed to do so. Another minor item is unconnected in any way to solving the puzzle, however. A British Reuters correspondent is suddenly released from a Chinese prison after the ending of prison sentences of some Communist journalists in Hong Kong. *Time* does not signify this move as presaging any changes in China's policy toward the West.

As if to serve as the capstone of a trend, *Time*'s last China picture of 1969 counterpoints a story that seems to predict the future: "China on the Verge of Speaking Terms." The narrative concerns the invitation of a Western diplomat to the Chinese embassy in Warsaw, the first such meeting, as noted above, since 1949. The diplomatic history of the United States in China is summarized in brief, and *Time* reports that "though the content of their discussion remains secret, President Nixon's top foreign policy advisers are convinced that Beijing may well be on the verge of resuming formal talks with the U.S." The story further reports that "the Nixon administration is anxious to draw China out of its 'angry, alienated shell' as Undersecretary of State Elliott Richardson put it recently." The United States' incentives to approach China include solving the Vietnam problem and deterring a Sino-Soviet war. China's motivation is described as basically "its fear of Russia."

Yet caution about any breakthrough is offered. *Time* notes, "For one thing, Chinese Communists demand, as a precondition for even the smallest agreement, that the U.S. abandon the Nationalist government on Taiwan. Also, few Westerners comprehend how far Mao's China will go to protect its ideological purity. . . . In fact, the most that the U.S. could hope for in the near future would be an agreement to hold regular discussions." Interestingly, the illustrations for this story include a picture of America's ambassador to Poland and a rather contrarian cartoon (see figure 5.1).

Figure 5.1. December 26, 1969: Cartoon, "Right Bait for the Right Fish?" (Reprinted from Wetzel—Ben Roth Agency.)

A caricature of Nixon, complete with his trademark ski-slope nose and tight business suit, sits on a sampan; his expression is preternaturally devious. Yet he extends a fishing pole on which the hooked line is an olive branch and the bait is a parchment reading "closer ties with Red China." Below the water, bulbously looking at Nixon (not at the bait) is not a Chinese dragon but a great fish with a shark's body and a round Mao-caricatured head with a tooth-filled mouth. The caption reads, "Right bait for the right fish?" *Time's* discussion of the possibilities of a Sino-American rapprochement indicates limits and, in cartoon form, anyway, the predatory and bestial view of Mao still lingers.

The year 1970 is another lean one for China images in *Time*, the calm before the visual explosion to come over the next three years. The first picture related to China is a humorous one, described above in the prologue, the "young priest" who is really Mao. Other coverage during the year focuses on Indochina and Beijing's relationship with different countries. A January 19 issue shows Zhou Enlai and America's ambassador to Poland, Walter Stoessel, and discusses the diplomatic triangle of the Soviet Union, China, and America. The story relates "Beijing's flirtation with Washington, however mild" to the tensions and conflict with the Soviet Union. *Time* quotes Harrison Salisbury as saying that the USSR, on its part, is "almost paranoid" that the United States might take China's side. Again, there is no speculation on any sort of real agreement between China and the United States Small daylight from the PRC is then offered in the form of another prisoner release, an elderly Catholic bishop; it is also reported that an American businessman, who had been held in China for spying, had committed suicide. Four other "non-Communist Americans" are listed as still being held in China. We can now detect a non-Lucian spin to this news: the death of an American is framed as the removal of some impediments to improving Sino-American relations.

In contrast, Taiwan, which has almost completely fallen off *Time* magazine's world map since the death of Luce, appears twice in 1970, once in a story about an assassination attempt against one of Chiang's sons while he was visiting America, and again in a photograph of a successful female Taiwanese Olympic track-and-field star. A story appearing in the October 26 issue is interesting in regard to what is about to occur. Canada had decided to establish diplomatic relations with China and "break off relations with Nationalist China, but Taipei beat [Canada] to the punch." The Nationalist ambassador left, severing ties with Canada. Other nations, like Italy, are described as also approaching recognition of the PRC. A photograph shows the ambassador and an unidentified woman, presumably a relative or a member of the embassy staff, in tears. Details follow on how China's diplomatic isolation after the Cultural Revolution is dissolving, with a reopening of embassies. The reasons for Canada's move are represented as economic—that is, the nation's growing trade with mainland China. Moral stanchion returns in *Time's* subcaption of the tearful Taiwanese woman, "Pleasing a good customer"; of course, in the decades to follow, good business relations will became a major reason for the United States to improve and maintain its ties to mainland China.

Time's final images of China for the sketchy year of 1970, published in the November 16 issue, include a belated photo-essay on the twenty-first anniversary of the founding of the People's Republic of China. *Time* emphasizes that the pageantry and spectacle that are shown in a number of photographs of demonstrations, scenes from the opera, parades, and pantomimes, for all their ideology tend to emphasize "practical 'socialist reconstruction.'" *Time* argues that China may be emerging from its "long night," which "began in 1958 when Mao launched his ill-fated Great Leap Forward." *Time* sees "a new awakening" for China in the nation's "emergence from an almost unfathomable period of mass self-flagellation." China's opening of embassies and moderate economic progress are cited. A sidebar story without illustration discusses the "pros and cons of recognition," detailing the number of nations that determined to recognize Communist China, most recently Italy. The story states that the United States is now the "chief lobbyist against China's admission to the United Nations." Another vote, whatever the result, "will increase pressure on Washington to modify its China policy." The pros and cons are then listed as bullet points, and the demarcation is interesting because it is without the heavy-handed pen of Henry Luce illustrating, arguing, and anecdoting the strength of the cons versus the weaknesses of the pros of recognition.

One concluding line in the piece, however, shows that the Nixon administration has kept its intentions secret. Speculating on the "two China plan," that is, to recognize Taiwan as an independent country, *Time* points out that neither Chiang Kai-shek nor Mao would approve of such a move, but "virtually all State Department Sinologists feel that a slow conversion to a 'two Chinas' policy is the only worthwhile course open to Washington. They have been advocating such a course for three years, and now the Nixon administration is quietly moving in that direction." We know now, of course, that the Nixon administration was moving in the opposite direction: a one China, indivisible policy, as articulated in the Shanghai communiqué that would spring from Nixon's China trip. One wonders here whether "virtually all State Department Sinologists" were lying to *Time* or projecting wishes rather than the actual intentions of their higher-ups—Kissinger and Nixon, after all, were not Sinologists—or whether *Time* had been caught in the oldest deception of Washington journalists, projecting the leak of one source into a whole department. More likely, Nixon's hedging tactics, as described above, were working well, much to the future embarrassment, although without admission of such, of *Time*'s expert China watchers.

The first months of *Time*'s China coverage of 1971 confirm the "surprise" hypothesis of Nixon's coming announcement. The magazine's January 11 issue features a story illustrated with a cartoon of a pig-jowled, squint-eyed Chairman Mao loading up on weapons that he has plundered from a nondescript chest. The story discusses the mixture of some official improvements in Sino-Soviet relations and rhetorical barbs exchanged between the two nations or their proxies. *Time* concludes, "What apparently bothered Moscow most was the fact that

China's trade with its former ideological allies has dropped from almost $3 billion in 1959 to less than $800 million. China's trade with Russia alone plunged from $2 billion to a puny $57 million in 1969. Meanwhile, trade with non-Communist countries grew from $1.3 billion to $3.1 billion." Another story, in the January 18 issue, contextualizes what is to come. A map and graphs and a photograph of "Marines shipping home" illustrate the "shrinking U.S. presence" in Asia, that is, the drawing down of troops in Vietnam but also in other territories such as the Philippines, Japan, and Korea. *Time*'s criticism is nonexistent, and Luce's absence is palpably felt when the magazine seems to treat with respect an administrative policy "genuinely determined to give the U.S. a substantially lower profile in Asia." Henry Luce wanted an American century in Asia, not a retreat by America's centurions.

In *Time*'s March 8 issue, a cartoon from the British newspaper the *Sun* shows some caricatured grim-faced Chinese soldiers setting off a gigantic bottle rocket. The cartoon accompanies a story about Chinese advances in making middle-range ballistic missiles. The caption of one soldier talking to another reads, "It scares me, too, but the point is it also scares the Americans." The story asserts that the Chinese nuclear force is still far behind—"twenty years" behind—that of Russia and the United States. In the issue of March 22, a picture is shown from a state visit of Zhou Enlai to Hanoi. He is quoted as pledging China's support for North Vietnam, but makes clear that, as *Time* interprets, "China will probably not intervene unless North Vietnam is invaded or seriously threatened, or the fighting in Laos approaches China's border." U.S. Secretary of State William Rogers states that the United States had no plans for such an attack, but "we don't exactly rule it out, either."

Time's April 19 story on China seems to foretell that something new might be about to dawn. Titled "A Quieter China in a Calmer Asia," the story details the decline of crises in most of the countries in Asia and China's pullback from adventurism against its neighbors. Better relations between China and other countries are described or foretold. *Time* notes, "In the past several months, a suddenly cooler China has been the catalyst of a number of gradual shifts that have been taking place in the mood and the manner of Asia's capitals. Last week's invitation of the U.S. table tennis team to visit China was an example of Beijing's new approach."

The next issue of *Time*, on April 26, unleashes visually on its cover the new approach and the new world. A group of smiling American men and women, in sneakers and with many holding cameras, stands on the Great Wall (see figure 5.2).

It is a classic example of how an aesthetically and narratively dull image can achieve iconic status simply because of the important event that it portrays and thus symbolizes.[96] When I showed this photograph without captioning or dat-

Figure 5.2. April 26, 1971: Time cover, "China: A Whole New Game." (Copyright 1971 by Time, Inc. Reprinted with permission.)

ing to a two-hundred-person class of my undergraduate students (in 2003) and asked them about its qualities, all dismissed it as some form of tourist snapshot. There were a few comments, however, about the hokey fashions; one student asked if the two women wearing kerchiefs were Muslims. But on April 26, 1971, *Time* was able to describe "U.S. ping-pong players at Great Wall" as illustrating "China: A whole new game," and to post a brazen announcement: "First color photos! Yanks in Beijing!" Inside, the feature article announces "The ping heard 'round the world." Zhou is shown shaking hands with a member of the American team. Other pictures chronicle team members visiting sites in the country, practicing, and generally being warmly or curiously greeted by Chinese crowds.

The story also details some of the foreign policy background to the micro-breakthrough. Nixon is described by *Time* as feeling that it is "too early to talk seriously" about the major questions dividing China and the United States. The magazine summarizes, "Most of the answers could not possibly become clear for a long time, but the world experienced the refreshing breaking of a dreary stalemate. Even if this break might bring new risks, they seem preferable to the old paralysis." It is inconceivable that Henry Luce would have approved of this last sentence, and the story heralds the major theme—and later criticism—of the coverage of Nixon's trip to China. Americans, so long used to negative news from China, thirsted for the good, even if it was largely excellent stage management for their benefit. For example, members of the U.S. team are shown walking by "applauding Chinese stand[ing] by sign proclaiming 'Welcome warmly the U.S. ping-pong delegation.'" Such an image, displayed in 1958, would probably have drawn a Lucian subcaption of something like, "Behind the smiles and applause, the terror and repression continue." Some negative history of China is recounted, however: an undated photo, presumably from a previous era, shows "Chinese boy shooting arrows at [caricature of] LBJ."

But commentary is almost completely free of questioning what is going on behind the images. A *Life* photographer, Frank Fischbeck, is quoted as saying, "It was a big, smiling campaign through a vast country at very high speed." Interestingly, Nixon's hedging strategy of continues to be effective. The president laments that "China continues in its determination to cast us in the devil's role. . . . Our modest efforts to prove otherwise have not reduced Beijing's doctrinaire enmity toward us." Speculation on the diplomatic outcome of better U.S.-China relations, however, enumerates various options for what these relations would look like. Of the three options outlined for Nixon in regard to the question of seating China in the UN, none are what he would actually take within a year. Nixon did not continue to oppose China's being seated, nor passively fail to oppose the unseating of Nationalist China, nor accept China's UN admission only on the precondition that the Nationalists would remain seated.

Time's May 3 issue focuses on "more signals" from China and provides more pictures to illustrate them. An anonymous Russian diplomat is quoted as saying, "Mao invites a bunch of your ping-pong players and Zhou offers them lemonade, rice cookies, and a trip to the Great Wall. Mao could not have made a better public relations move even if he had denounced his own sayings and told the world he was Mr. Henry Ford's secret business partner. This is not foreign policy; it just shows that Mr. Mao knows something about Madison Avenue." But again the hedging strategy appears. Vice President Spiro Agnew is cited as being in opposition to a too-fast move toward responding to Beijing's opening. But then *Time* notes that Agnew did not get approval even from conservatives for his opinions. Pictures show street scenes, countryside, tourist poses, junks on the river—they are all captioned without any oppositional carping. *Time* does, however, conclude the issue with a story on "what they saw—and didn't see." The commentary, while expressing some skepticism, is denuded from the aggressive questioning of "Red propaganda" that we might expect from a previous age. The direct comments are worth quoting in full, for although the story notes "Big Brother is watching," and that materialism is not as completely rejected as the visitors may have thought:

> [T]he first rush of impressions brought or cabled home to the U.S. from China last week evoked an image of a society unusually unified and content within itself. The Chinese people seemed genuinely enthusiastic about their condition. With an almost disconcerting unanimity, they answered questions with an appropriate quotation from Chairman Mao Zedong. The image was undoubtedly too simple, though roughly true as far as it went. Still, it must be remembered that the travelers were shown mostly showcase spots that are on the itinerary of nearly every foreign visitor. As fascinating as those sights were, they hardly gave a full view of Chinese society. A somewhat more accurate picture emerges in comparing the American visitors' impressions with those of earlier travelers and China watchers able to point out what the visitors did not see.

Further, *Time* concludes that in this "Poor Man's Paradise the secret of Mao's China can perhaps be summed up in an old Chinese saying: 'the contented man, though poor, is happy; the discontented man, though rich, is sad.' One reason why the average Chinese appears happy is that the wide disparities of wealth that lasted into the 1950s have disappeared." This is an incredible series of conclusions, not incredible in the sense of not being possible, but they would certainly not have been part of *Time's* analysis only a few years earlier. *Time* is describing a China that many of its enemies, the old China hands, had long imagined. This is not to say that the newsmagazine was romanticizing China; the conditions in 1971 *were* different from those in 1967, 1962, and 1958. But *Time* basically accepts the vision of China projected by Mao, its founder's old nemesis. Was this a sign of the collapse of the Cold War paradigm? I argue that a psycho-social force rather than an ideological one was at work here. Americans wanted a new image of China, and when it arrived they were willing to embrace it, despite artifices and flaws.

Photographs in the weeks to follow set the stage for something to happen, but we are not sure what. In the June 7 issue of *Time* is an article on the future of Taiwan. It is as dispassionate has *Time* has ever been, revealing its new paradigm of disengagement from China as an issue of emotional attachment or as part of an anti-Communist crusade. The future of the island is indeterminate and "not so easily settled." Various policy options are offered, but again *Time* is waving no flags. Other China-related stories in the weeks to follow deal with Indochina and India, but in *Time*'s July 12 issue, dated a few days before Nixon's sensational announcement about his intended trip, a feature story on "Mao's Attempt to Remake Man" contains no hint of what is about to transpire. The description is once again unheralded from the previous editorial view of *Time*. The story is not completely positive. A little boy, a "tot with Mao's book of thoughts," is subcaptioned as "learning to hate the old," but there is definitely an awe contained in the fact that "merely gaining effective control over China's eight hundred million people . . . was an epic achievement." Mao's remaking of the Chinese individual "into a new Maoist man" is described as largely successful. The descriptions of mass indoctrinations are clinical and tinged, when coming from the mouths of university professors, with fascinated description. One senses a *Time* that doesn't quite know what to make of China and that can't bring itself to find fault in the spectacle of the Chinese mass.

Time's July 26 issue's cover reads, "To Beijing for Peace." In a jovial caricature of Washington's crossing the Delaware, a cartoon Nixon stands at the prow of a rowboat decorated with an American-flag motif. A floral suitcase is at his feet. The oars of the boat are falling aside because a Henry Kissinger with thick eyeglasses mans a furiously belching and splashing motorboat engine. In the distance is a Chinese pagoda flying the PRC flag. The metaphor is an obvious one: Nixon is crossing the great gulf between the nations by jump-starting what was supposed to have been a painful process of chasm-crossing. Inside the issue, pictures show Kissinger and Zhou in Beijing shaking hands and Nixon greeting his aide "on return from [the] secret journey." The story details "Nixon's coup to Beijing for peace," describing Kissinger's trip and the plans for Nixon's China odyssey. *Time* notes, "It would be ironic—and yet appropriate—if the man who launched a political career largely on the basis of his fervent anti-Communism were to cap it by establishing himself as a leader who helped move the capitalist and Communist worlds toward a historic rapprochement." The story does caution, and we know now that this was the press line of the White House, that "to be sure, much could still go wrong. There is no certainty that any of the high expectations will be achieved. Obviously, unforeseen events could prevent" Nixon's trip.

The response to the news is described as "highly favorable" across the board, except for "only a few conservatives." William Buckley Jr. is quoted as saying, "FDR would have hesitated to go to Berlin to wine and dine with Adolf Hitler—but we are about to do that, and all the liberals who can't stand the Greek colonels [of that country's pro-American dictatorship] are jumping for joy." The pictorial illustration of the people's reaction negates such carping. Nixon is shown mobbed by a multiracial group of individuals at a restaurant "after trip announcement." The subcaption reads, "The unpresidential elation was forgivable." Nixon is rumple-suited but smiling broadly, as are the people. The story is almost dismissive of the danger to Taiwan: "The Chinese are not likely to attack Taiwan anyway; any such attempt would be a bloody and costly venture." Yet "the long-range future of a non-Communist Chinese government on Taiwan is not bright."

Time's August 2 issue continues with a somewhat chastened and chastening story on "Hazards along the Road to Beijing." An Oliphant cartoon shows Nixon carrying a suitcase marked "China," tiptoeing, trying to hide his face from a giant portrait of "John Foster D" (Dulles) who is almost bursting out of the picture in astonishment. A cartoon from an Israeli newspaper shows Nixon being helped up the Great Wall by Mao; the guard tower, which is also a voting box, is labeled "Election '72." Other images in the issue show some anti-China protesters, and again the story reports the only American opposition being "scattered voices on the far right." *Time* speculates on what will be discussed during the trip and the secrecy surrounding the substance of the "arrangements, agenda, or policy implications of the summit." An interesting sidebar is given to Edgar Snow on "Why Nixon is Relatively Good." The illustration is the famous photograph of the journalist and Mao standing together at the Tiananmen podium. Snow was, we recall, not a favorite of Luce, but here he is described as having "had unequaled access to the thinking and policy shifts within the Chinese government." Snow confirms Mao's predilection for dependable enemies, saying that Mao told him, "He preferred men like Nixon to social democrats and revisionists, those who professed to be one thing, but in power behaved quite otherwise. The Chinese believe that the lesson of Vietnam, and no mere change of presidents, is what made it possible for Mao in 1970 to speak differently about Nixon. 'Experience' had made Nixon relatively 'good.'" A following story, "Meanwhile in Taiwan . . . ," discusses the misgivings of the Nationalist leadership, but notes how Chiang himself has "kept his dignified cool." The story concludes, "The Nationalist leaders feel that all is not yet lost. . . . Hope is not much to cling to, but it is all the Nationalist Chinese have." Finally, there is a picture of John Service and John Paton Davies about to testify in a Senate hearing. This event marked the flowering of the "lost chance in China" school of argumentation, and it is fitting that it is noted by a picture in *Time* magazine.

In *Time's* August 9 issue, Zhou is shown sitting down with a visiting American student from the "Committee of Concerned Asia Scholars," which was and is a leftist academic group. The Chinese dragon returns in a cartoon from the *Dayton Journal*: a huge, fat beast, sporting a sleepy Mao face, marked "China," with a giant hammer and sickle, squints its one open eye at a small businessman carrying a briefcase marked "Japan's economic growth." Another story contains illustrations on Chinese acupuncture practices. *Time's* August 16 issue covers more of the reaction of the Taiwanese to Nixon's upcoming China trip. One photograph shows "students marching in Taipei" holding protest placards and Taiwanese flags. A cartoon shows a bewildered Chiang as a lion tamer fending off, with a small chair labeled "U.N. seat," a giant lion whose maw is labeled "China." A further story on the upcoming United Nations vote describes how America's position to retain membership for Taiwan as well as admit Beijing was eroding. George H. W. Bush, then UN ambassador, is "having great difficulty lining up support for this position." A cartoon shows a UN workshop craftsman trying to design a China chair with split seats.

Time's October 4 issue strikes the first real negative note about China since the U.S. ping-pong team's visit. A story titled "China: Signs of Internal Strife" speculates on the cancellation of the scheduled parades and spectacles for National Day on October 1. Lin Biao, Mao's heir-designate, is missing—it is questioned whether he may be ill or dead. The story also asks whether it might be Mao who had died. The enigma continues in the October 11 issue, with more speculation about the missing leadership, although Zhou Enlai made an official appearance, meeting a Cambodian delegation. He is the only high-level leader, *Time* reports, who "has been in the public nearly every day."

Then *Time* features Zhou on its cover of November 8, 1971. It is an unflattering pencil drawing in which Zhou performs the customary one-fifth-turn "visionary stare." He is expressionless; the banner headline reads, "The Chinese Are Coming." The inside story details the vote in the UN that recognized the People's Republic of China and withdrew recognition of Taiwan. Scenes of protest and congratulations accompany a photograph of the Nationalist foreign minister leaving the UN building, head bowed. A final story illustrates an image not long seen in *Time*: Chiang and Mao stand together, their last meeting at "U.S.-backed talks in 1945." The story expresses a "future uncertain" for "Chiang's last redoubt."

Other elements appear in the story that were atypical of *Time*: "Politically, as the balance of world recognition continues shifting from Chiang to Mao, it becomes doubtful that two million mainland refugees can continue their authoritarian rule over twelve million Taiwanese." The story speculates, "When the elder Chiang departs, even reunification is not inconceivable." Finally, some of the last China pictures of the year, in the November 22 issue, detail "the fall of Mao's heir." The story speculates that Lin Biao has died in a plane crash (which was correct) after being purged (also correct). The circumstances are mysterious. Zhou Enlai, though, is pictured as coming to power as Mao's new heir, and in all the stories he is consistently painted as being prorapprochement with America, a relative moderate. In the November 29 *Time*, the new PRC delegation to the UN is shown laughing in their seats. Among them is Huang Hua, the Chinese diplomat who negotiated with Ambassador Stuart in the last days of America in China in 1949. The only visions of China in the last three issues of 1971 relate simply to China being on the map in stories about the India-Pakistan war.

Among *Time's* pictorial retrospective of 1970 in the January 3, 1972, issue are the cover photograph of the grinning U.S. ping-pong team at the Great Wall; the People's Republic of China's delegation to the UN laughing with joy at their new seat; and a photograph of Kissinger and his party at the Great Wall. Inside the issue is a story of the "ouster at the U.N." of two correspondents from a Taiwanese government-owned newspaper. *Time's* editorial solidarity with Taiwan has now vanished; although it states, "The rationale was plainly political and discriminatory," the magazine is expressing indignation on behalf of fellow journalists, not of Chiang's regime. A story a week later shows medical news from China, with images of "barefoot doctors," medical instruction for peasants, and a somewhat gruesome image (to the Western eye) of a "child undergoing acupuncture." But the story itself reads like a press release by the Chinese government: "The Communist regime made an early policy decision to marshal all medical manpower for two basic missions: to make some sort of care available to the entire population and to drill the people into becoming active participants in public health campaigns." The story even contains some positive references to the Cultural Revolution and the Red Guards. The "Red Guard doctors" in the cities are described as community health workers; "dispersing health care throughout the country took on momentum when the Cultural Revolution began in 1966."

The two most copious weeks of China images of the entire survey to date begin with *Time's* issue of February 21, 1972. A cartoon transmographies the Chinese mass into a popular culture incarnation of '70s Americana: caricatured Nixon and Mao shake hands in a Chinese pagoda. Below them is a sea of variously slant-eyed and round-eyed smiley faces (see figure 5.3).

The interior section prepares America through "A guide to Nixon's China journey." Portraits of Mao and Nixon are on the same page as a "group of smiling, applauding citizens on crowded public transport bus in Shanghai." We

Figure 5.3. February 21, 1972: Time *cartoon, Nixon and Mao and Smiley Faces. (Artist: Mike Smollen. Copyright 1972 by Time, Inc. Reprinted with permission.)*

presume that they are smiling at and applauding the news photographer. The story tells us, "The Beijing summit fairly shimmers with the kind of historic aura that Richard Nixon dearly treasures—the leader of the world's most powerful nation meeting with the ruler of the most populous. . . . Even if nothing happens at their meeting—and no dramatic breakthrough is in sight—the reopening of a U.S.-China dialogue has fundamentally altered the power structure of the globe." This description is interesting when it is contrasted with some of the postsummit analysis where, indeed, "that nothing happen[ed]" will be the subject of some degree of remonstrance. In short, here *Time* sets up high expectations, at the same time lowering them, demonstrating the success of Nixon's hedging strategy; then, as we shall see, the magazine gets carried away in its adulation only to unselfconsciously sober up later. The story continues,

> The ceremonial portions of the seven-day visit will be televised live by satellite to a worldwide audience that may match or exceed the estimated six hundred million who saw man's first steps on the moon. . . . The Chinese are clearly going out of their way to make Nixon's visit pleasant and untroubled. On the eve of the journey, China is undergoing a great wave of repainting, face-scrubbing, and sign-switching. Slogans referring bluntly to "U.S. imperialists and their running dogs" have been replaced by blander remarks about "the unity of the world's peoples."

Nixon's hedging is repeated; the White House, we are told, "does not expect anything large or sensational," and Nixon and Kissinger "have stressed [that] no deals affecting third countries will be made in Beijing" (this is a reference to Taiwan and South Vietnam). Photographs, maps, and sidebars describe the different cities and locations on the Nixon itinerary. An article features Zhou as "the man in charge."

Again, it is a different *Time* magazine that tells us that Zhou, once a killer (according to *Time*), is now the "regime's suave spokesman and negotiator." An article on "Life in the Middle Kingdom" is illustrated by a picture from Eastfoto of "Guards on Tibetan frontier visiting villagers." It is not clear whether this is the frontier of China with Tibet or the frontier of Chinese-occupied Tibet. That the story makes no remark about the difference shows how much *Time* has changed, as well as the times. Another picture that shows "Peasants us[ing] oxen to plow land used in commune" would have been a choice illustration of a previous-era *Time*'s point about the failures of Chinese communization as well as Communism. Now, the story tells us, "Westerners who remember the pre-1949 China . . . have been almost euphorically impressed by the transformation that Communism has achieved." Nevertheless, *Time*'s report underscores that this is a *Chinese* achievement, for "however pleasing its surface appears, China's 'future' is one that most Americans would find intolerable." The story concludes that although the Chinese Revolution has progressed "at a terrible cost," still Mao's single-minded struggle to transform China "seems to have been successful."

In the February 28 issue, *Time* pays a tribute to "China expert Edgar Snow," shown smiling in a photograph subcaptioned "Lifelong love affair." The story details Snow's health problems and later his death, tragically "just six days before President Nixon's arrival in Beijing." The story continues, "Mao, in a personal message to Snow's widow [said], 'His memory will live forever in the hearts of the Chinese people.'" The accompanying biography of Snow mentions his book *Red Star over China* as a "masterpiece of reporting"; the story does discuss the various charges against him, among them that he overly romanticized the Communists, but admits that he was able to critically assess their failures as well. It would be hard to imagine that a Lucepress obituary would conclude, "Snow's reporting remained indispensable to serious students of China." John Hamilton, in his biography of Snow, notes how Nixon's trip was immediately framed as a vindication of Snow's persistent contact with the Chinese and his attempt to debestialize the image of Mao. Because Snow, like many "experts" on any subject, wrote in great volume in great range over a great time, he also served, as Hamilton points out, as a sort of "quote book" for anyone holding any opinion about the People's Republic of China. Within Snow's corpus we can find praise of Mao and condemnation; we can see both the blemishes and the virtues of the decades of Communism in China. In short, one of the most important image makers of China who contributed, for example, the most famous photograph of Mao (in front of the Yan'an caves—as we recall, wearing Snow's borrowed cap) became himself an icon, still part of the debate about the nation and the people that he loved.

Time's March 6, 1972, issue is almost completely devoted to Nixon's trip to the People's Republic. So photo op saturated is the event that *Time* provides an unusual split cover containing four images instead of the usual one: Nixon shaking hands with Mao; Zhou inviting Nixon to sit; the Nixons traversing the Great Wall; and a photograph from a scene of a musical from the Red Opera. The banner headline is "Nixon's China Odyssey." An interior note tells us that the quadpartite division of the cover has its images separated by "a stylization of the Chinese character for friendship." Inside the issue, the letter from the publisher, Henry Luce III (son of the founder), writes, "'There is a kaleidoscope quality about the events and the trip.' White House correspondent Jerrold Schechter cabled from Beijing last week. 'Each moment fixed, then whirling on to a new sensation. This has been a week of sights and sounds.'" Luce then goes on to describe how the journalists have tried "to capture the quality of life in a country few U.S. journalists have seen in more than two decades." Images are foregrounded in this description of the work of producing the coverage. Luce continues, "Tracking their every move from over six thousand miles away was picture editor John Durniak. Since the day the Nixon trip was announced, Durniak had been laying plans to get photographs out of China and into this week's issue. After equipping [the *Time* journalists] with cameras, he instigated a special airlift to get pictures of the trip off the mainland, and by Thursday night the first 150 rolls of film had been flown into Chicago. There, [the *Time* staff] worked nonstop for the next thirty-eight hours. . . . The result: a hard-won look into a long-hidden China."

Time's description of the event is largely adulatory, although with some reservation. It reports a favorite joke in Washington: "The trouble with Chinese diplomacy is that an hour later you are hungry again." *Time* asks, "What, if anything, did Richard Nixon bring back from Beijing? Above all, the event itself, the fact that it took place. . . . As summits go, the meeting was a glittering technical success, stage-managed with precision. . . . Nixon comported himself with dignity but with an enthusiasm that sometimes made him seem overeager, said nothing of importance in public during the entire trip. [There were] no great surprises, no great letdowns." The photographs are much more unalloyed in their praise. Nixon and Kissinger sit in armchairs with Mao and Zhou; the Nixons smile broadly at the Great Wall; "A scene from the Beijing Ballet Troupe's [*sic*] 'Red detachment of women'"; Zhou and Nixon talking at a conference table; Mrs. Nixon posing with Chinese women watching a Chinese cook at work and admiring "a bronze dragon statue at the summer palace." There is also a close-up of a giant panda, which, as said, and perhaps on purpose, would soon replace the dragon as the dominant symbol of China.

It is indeed a very touristy montage, full of diplo-routine, remarkable only because it *is* China. A full-page photo spread details in four images Zhou and Nixon toasting each other. A Chinese soldier raises the American flag, with a portrait of Mao in the background. The Nixons inspect a giant stone elephant at the Ming Tombs. An essay on "the president's odyssey day by day" features a Chinese man holding up the *People's Daily* issue that shows Nixon and Mao meeting. Other images show "Snow clean-up in Beijing" by Chinese women; "Nixon helps Zhou take off [his] coat"; Pat Nixon taste testing Chinese food and feeding pigs at "people's commune"; Nixon shaking hands with a Chinese boy. The narrative is not as praising of the president as the photos would indicate; the reporters write, of the Great Wall, "Even Richard Nixon's banal description of its might fails to mute the wonder of the morning." Another notation reads, "The President and Mrs. Nixon seem more interested in posing for pictures than in actually walking on the Wall." Here we have no doubt the resentment of the word journalist who is stuck for one week having to write descriptions of someone else's family vacation. At the time, as we have seen, Nixon's trip was designed for the cameras. The substantive issues between America and China could not be solved immediately, and the images were in fact the most potent diplomacy possible, to let both peoples see that their respective leaders could get along. The details would follow later, and would indeed matter.

Further pictures show schools, shopping, soldiers, and other scenes from China. There are a number of comments on the event *as* an event. For one of the first times in its history, *Time* shows pictures from television: Barbara Walters interviewing a Chinese interpreter, and famous journalists including Theodore White and Walter Cronkite eating a Chinese meal with chopsticks. The story comments, "For the television viewer, the president's trip was a remarkable demonstration both of TV's powers and limitations. No written account could convey, as did the live camera, the drama of Nixon and Zhou touching glasses after a quarter-century of enmity. At the same time, no written account could be as tedious as a camera searching for something—almost anything—to record." While many iconic images were produced, *Time* notes, "the thrill of discovery quickly wore off. TV crews and reporters were soon scurrying frantically to satisfy the medium's insatiable appetite for novelty, sometimes achieving massive inanity instead." The story also maintains, "Print journalists complained that they were alone while the TV reporters came in in teams with large supporting staffs. Some complaints seemed somewhat petty: TV network staffers were provided with cars, while writing journalists had to use buses." Also noted was that, "Deprived of the customary briefings and backgrounders, correspondents were forced to fall back on color and trivia." A series of cartoons pokes fun at the visit: Nixon floating off his airplane at the Beijing airport surrounded by doves holding an olive branch, and a briefcase inscribed "I am the one," all topped off by a banner proclaiming: "Peace for all mankind" (see figure 5.4).

Another cartoon shows Nixon boring Zhou with details of Chinese history (fed to him by Kissinger). A third shows Nixon and Mao surrounded by a huge crowd of journalists, onlookers, cannons, airplanes, and photographers; Mao whispers to Nixon, "Just think, all this will have gone to waste if you're not reelected!"

Time's March 13 issue reports that Nixon's journey to Beijing was setting off a "sudden American appetite for things Chinese." The story relates, with some humor, "Across the nation, there was a great fumbling and clicking of chopsticks—an item that restaurants often ran out of, as Americans accustomed to forks and chop suey suddenly demanded authenticity." (This is in itself an irony, unnoted by *Time*, since chop suey is a wholly American invention, unknown in China.) Other evidence: "do-it-yourself acupuncture kits" selling at the Harvard Co-op. The illustration is an ad for Delta Airlines that is written completely in Chinese character script. The characters do not refer to trips to China but are simply an attention-getting device; a small note within the ad reads, "If your Chinese isn't as good as your English, see translation on page 23." A series of cartoons in the next story on the "Descent from the Summit" details the "somewhat

Figure 5.4. March 6, 1972: Cartoon, "It's Going to be a Long Week . . . /Peace for All Mankind." (Reprinted from Dayton Daily News. *Artist: Mike Peters. Courtesy of Dayton Newspapers, Inc.)*

anti-climactic . . . letdown" following the visit. *Time* in its reporting here uses a typical device of journalism of projecting false emotions on others; the story describes, "It is generally expected of great historic events these days that they should have some immediate, dramatic consequences. No one could deny that Richard Nixon's Beijing summit was such a historic moment, or that the public's expectations were intensified by its being the first event of such magnitude ever to be staged for television." Of course, as we have seen, *Time* itself contributed to such a magnification and, we may also say, glorification of the summit. If there is an anticlimax, it is journalism's anticlimax, perhaps the feeling of a Chinese buffet-goer who has indulged too much in the "all-you-can-eat" offerings or, in this case, the all-you-can-photograph spectacle.

A series of cartoons pokes sometimes gentle, sometimes more acerbic humor at the events. The most grim is a Mandarin-clad Nixon, incongruously with a huge Arabic sword at his side, bowing to Chairman Mao, who bows back. The sword has flown backward in the air and splits a sad figure of Chiang Kai-shek in half. Three other cartoons poke fun at the Sinoization of the presidency: Nixon speaks to a television camera from his desk with his fingers pulling his eyes back to make them look slanted; his wife and a crowd of aides and cabinet members gather with a sign reading "Welcome back, Mr. President," but all are dressed in Mao suits; a workman takes down a "The White House" sign from the gates of the building and informs a passerby, "He's changing the name to 'The House of heavenly peace and tranquility and reelection'!" The opposition to the odyssey is still pictured as "just [that of] some conservatives." The Taiwan issue is raised, but Kissinger is described as insisting, as the story puts it, "Appearances to the contrary, the U.S. had not given anything away." There is some discussion about the "Potemkin village" atmosphere of what was pictured, but it is hardly a negative criticism: the Chinese "may, in fact, have been a little too attentive to detail. After American reporters discovered that the well-dressed, cheerful Chinese milling about the Ming tombs with transistor radios had been planted there to impress the visitors, [Zhou] told Nixon by way of apology: 'We don't claim to be perfect; we shouldn't have done it.'"

Stories in the weeks to follow focus on Indochina and cultural pieces about the mainland. A photo-essay on April 10 details a "reporter's second looks." The story is of *Time* correspondent Jerrold Schechter, who was "one of two U.S. newsmen who were allowed to remain in China after the Beijing summit ended [and] took a twelve-day journey through the land of Mao." It is an almost completely laudatory essay; the pictures are nearly all those that one might be expected to see produced by the propaganda bureau of Eastfoto. Almost all, we can also imagine, would have been oppositionally captioned in a 1950s issue of *Time*. They include "youths boating on Nanjing Lake"; "auto workers with Shanghai sedan"; "typesetters at [newspaper]"; "Nanjing militia pouring steel to make harrow blades at foundry"; "girls operating welding machine at Shanghai shipyard"; and a portrait of a "Chinese woman factory worker."

There are other pictures that, however, the Western viewer could scarcely view positively, although this era's *Time* feels no need to emphasize that sociopolitical gulf. How should we react to a picture of ribbon-haired Chinese girls holding huge rifles "at target practice"? Also, in another era, a photograph of "team of commune workers pound[ing] the earth for Yangtze River irrigation project" would surely have occasioned a contrapuntal caption about famines and forced labor. The story itself describes China as a land of egalitarianism, but the reporter was obviously disappointed by the aesthetic and intellectual monotony that such a society presents: "In three weeks, I did not see a single woman wearing a skirt or a dress." Western journalists in China, the story implies, do not have much freedom to be Western-style journalists. Schechter concludes the story by noting the difference between Moscow and Beijing: "China has a sense of vitality that is not found in the Soviet Union. Russia, which I toured extensively as *Time*'s Moscow correspondent from 1968 to 1970, is a classless society with a privileged elite. In China, by contrast, everyone is poor together." Although Schechter does say that this situation produces a "drab, intense, and self-absorbed society," there is an incredible omission here. Schechter, like the other

American journalists, had spent two weeks covering banquets and parties for the American and Chinese elite. We certainly know, from Mao's physician's memoirs, that the members of the inner circle of Chinese leadership, the so-called Group One, never lacked food, comfort, or merriment. Mao also had his own cadre of teenage peasant girls available to gratify his libido. In short, China had a privileged elite. One could say that the American journalists missed those privileges because China in 1972 was a much more closed society than the Soviet Union but, again, we note that Schechter had just spent weeks indulging, along with President Nixon, in those very privileges.

The pictures for the rest of the year 1972 deal with Chinese diplomacy, acupuncture, and another trip to China with a Western reporter following a Canadian delegation; the illustration of the last is a photograph of a busy "snack stand near Beijing's Workers' Stadium." The last major China story of the year is in November 20's *Time*, when another correspondent toured the country and discovered "the dividends of rediscovery of China . . . is a continuing journalistic event." More photographs show the quaint and the cute, in street scenes and with a smiling "Beijing to Shanghai stewardess."

The post-China trip, postelection year of 1973 begins in *Time*'s January 1 issue with a cartoon and photographic retrospective of the president's tenure, which is in no way fully laudatory. In fact, the corrective against Nixon spending too much time on foreign policy is visually and verbally evident. On one page are photographs of Nixon and Zhou toasting each other; Nixon sharing another toast and chat with Leonid Brezhnev, premier of the Soviet Union; and above all yet another picture, which, if it were not for the subjects in the image, its relation to a particular event, and the multidecade metaphor it once again shape shifted, would be classified as a dull tourist snapshot: Nixon and his wife, Pat, on a tour of the Forbidden City, smile broadly at a giant bronze dragon (see figure 5.5).*

The dragon is fierce and fanged, its head scales standing on end; it is meant to ward off evil. But Nixon, in long coat and tie, and Mrs. Nixon in mink, are laughing uproariously. Only the most churlish Nixon hater would not appreciate the president's sheer delight at the moment and the spectacle. This is what, in essence, the China venture has come to: the dragon, so often used to stand for a menacing China, is no longer mysterious, and certainly no longer threatening or terrifying. It is an ornament, an object of touristic titters. I myself, when visiting the Forbidden City for the first time on an August day in the early 1990s (along with what seemed like thousands of other tourists), regarded this dragon, or one that looked much like it, and I watched the long line of Americans, Belgians, and even Taiwanese and Japanese register it complacently, perhaps stand

* Note: Picture was actually taken on February 25, 1972, but not printed in *Time* until a later date.

Figure 5.5. January 1, 1973: Photo, Nixons on a Tour of the Forbidden City Looking at Chinese Dragon Statue. (Dick Halstead, UPI/Corbis)

for a posed photograph, and move on. The metaphor of the Chinese dragon was not dead but it had become, as soon everything would be in China, commercialized.

On the facing page, however, a series of cartoons snipes at the trip, even though Kissinger and Nixon were voted *Time*'s "Men of the Year." These are not critiques of the oversaturation, romanticism, or turnabout of the China image. Instead, they harp on the following themes: First, where was the peace dividend? A cartoon referring to Nixon and Kissinger's famous dinner at Perrino's restaurant celebrating Kissinger's return from his initial trip to China shows the presidential adviser at a liveried dining table with a laureled woman in white labeled "Peace." The cartoon reads, "The president thought it would be nice for us to be seen together." A second critique was that the China coup was really due to staff work—that is, Kissinger—rather than presidential initiative. This may be a fall-out of the president's "hedging" strategy. Nixon is shown stepping off a plane with Kissinger behind in the doorway. A crowd of capped Chinese wait below. One says to another, "Who's the guy with Kissinger?" Finally, referring to a theme of the accompanying story about Nixon's remoteness, his failure to reach out either to the press, Congress, or the American people, a cartoon shows Kissinger in an armchair speaking to another man in an armchair whose face and body are not visible but who we presume is Nixon. Kissinger says, "Dick, we visited the people of China and Russia; maybe it's time we visited the people of America."

Coverage in the weeks to follow focuses on the Indochina war, with China featured in maps of the area. Nixon is shown in a cartoon on the January 29 cover, a mock-up of the swearing-in ceremony for his second term as president. A banner reads, "Nixon II: Beyond Vietnam." The crowd in attendance, however, includes Leonid Brezhnev, Zhou Enlai, and a host of other representatives of American and world factions, including a protester and a POW. In short, while China is now one of the "problems" facing America and the reelected president, it is not a monster—Zhou seems downright friendly. A turnabout in policy is further implied in the February 12 *Time*, when a story on "entering an uncertain age" deals with how mainland China no longer wants America to withdraw from Asia but, instead, "now hint[s] that Beijing would be pleased to see the U.S. keep its bombers in Thailand and the Seventh Fleet in Asian waters for the time being." Visually, peace feelers continue. In a story connected to the return of American POWs and then Kissinger's trip to Hanoi to negotiate reconstruction aid to Vietnam, Kissinger and Zhou are shown once again smiling and shaking hands. An old friend of Chiang and a flayer of Communist China, Joseph Alsop, smiles from another picture at the Great Wall, now the symbol of "being there" in China—as it is for every tourist to this day.

China becomes the focus of more "normal" reporting in the weeks to follow: street scenes and everyday life, such as people in Shanghai practicing tai chi. A portrait of Mao in the April 16 issue mentions that he is now consulting the philosophy of Confucius. One minor image attached to a story that will prove much more significant later is a portrait of Deng Xiaoping, smiling and clapping; it is subcaptioned "low-key return." Deng, who had been twice purged, is described as "returned from the shadows that had enveloped him since 1966." The story tells us, "Deng's duties are modest: he is believed to have been put to work reorganizing the youth corps." Human interest and diplomacy mark other pictures of China: a visit from an American actress; a high-ranking diplomat appointed to

China as envoy and ambassador in all but title. In an almost quaint story in the July 2 issue, there is a depiction of the "intrepid moles of Jinmen." The island still stands, but the "reconquer the mainland" signs and the defenses are described as more like a waxwork museum of the Cold War than a threat to world peace. As the story describes, fighting is reduced to "odd-numbered nights [with] forty or fifty symbolic rounds [exchanged] which explode in sparsely-populated areas and cause little damage."

Other China imagery of the year includes a traveling exhibit of dynastic art from Beijing to Europe. We recall that such foci on dynastic art during the Luce-colored '50s tended to ideologically contrast the glories of Chinese civilization with the Communist iconoclasm. Here, it is straight arts reporting, with no political commentary except that this is a demonstration of China's opening up. There are signs of trouble, however. On September 3, a picture of an unhappy-looking Zhou is juxtaposed with a portrait of the philosopher Confucius. The accompanying story notes that Confucius had lately been attacked in the Chinese press, and this may have been a veiled reference to the leadership aimed at Zhou Enlai. In the issue of September 10, though, Zhou is shown next to a smiling Mao at a party congress, his status reconfirmed as China is "putting its house in order." But there are also hints of more struggles. A newcomer, leader of the Red Guards of Shanghai, Wang Hongwen, has been appointed to a high post, and is featured in two pictures.

The final China-related image of the year 1973 in *Time* is found in the November 26 issue. Once again, Kissinger is visiting China and meets with Zhou. The story tells us that the "prime purpose of the latest get-together was to discuss further improvement of relations rather than inaugurate any new policies." The road to normalization is the main issue, and Kissinger assures the Chinese that the United States is committed to resolving all of the related concerns. One major notation by Kissinger, however, is incorrect, and in fact will suppress the next four years of China policy. Kissinger promises the Chinese that Watergate will not divert China as a priority. It will, and true rapprochement and full mutual recognition and normal relations with China would have to await a new presidency and a new era.

Postscript: Normalcy and Disaster

When Henry Luce spoke of the "American century," he projected a global phenomenon, a nation whose affairs knew no borders and whose model anyone could and should emulate. The great publisher and the policymakers of Washington, however, assumed that some situations and localities were more deserving foci for the projection of American power than others. From the 1930s onward, Luce himself chose China; he was concerned about Europe—and Rhodesia, for that matter—but in his estimation, the most important region in deciding the fate of the world was East Asia. The World War II allies, with their official "Europe first" (and unofficial "China last") grand strategy, thwarted such a predilection. In the global struggle against Communism, China would rise in importance gradually as American policymakers applied "containment" to both Europe and the East. But the perimeter of the Cold War was still fluid: Korea was a footnote in 1949 but a spear point in 1950. In contrast, defending Tibet was not worth a war against China, but the United States did assist its rebels, until rapprochement with their occupier closed the relationship. Indochina posed a greater puzzle, with the primary question being: *Fight where to defend what line?*

In the same sprit, John F. Kennedy promised in his inaugural address that the United States would "pay any price, bear any burden, meet any hardship, support any friend, oppose any foe." It is not difficult to assume that a defense of West Germany and South Vietnam were equally in his mind. But after the Korean War, hesitation at fighting with China on the mainland or on its borders became enshrined as a foreign policy dictum. As we have seen, Sinophobia drove America into Vietnam but then placed unsupportable limits on the rules of engagement. Kissinger and Nixon were of the first administration to argue (and admit), however, that America could no longer support (in money or lives) a policing role as a dominant superpower. Gone would be—as Kissinger hoped—a situation that had existed since the late '40s of "two rigid alliances with very little diplomatic maneuvering room between them [who] faced each other . . . all around the globe."[97]

China in 1972, in the eyes of Nixon and most policy analysts, was no longer a menace, nor was it a member of any "rigid alliances." Taiwan was safe behind the Seventh Fleet. The shelling of the offshore islands had become less Gilbert and Sullivan and more like a court ritual, with propaganda leaflets rather than explosives shot across Amoy Bay. America wanted to abandon Southeast Asia but needed to salvage at least the formality of "peace with honor." And the two polarities of Communism were at each other's throats. But would reaching out to one produce better relations with the other, as Washington hoped? Kissinger felt that the animosity between the Russians and the Chinese was so strong that neither had any "real alternative to us as a counterweight" and so we in America could have a "Mao tai and drink our vodka too."[98] But neither antagonist appreciated being a "card" to play against the other. From the Chinese perspective, rapprochement with the United States had not gained them an ally against Moscow. In fact, they watched with irritation as détente thrived. Kissinger repeatedly met with Russian leaders in summits and signed major pacts including SALT I, trade treaties, and the Helsinki Accords. The CCP leaders would push Kissinger, as he described in his memoirs, to "take the lead in organizing an anti-Soviet coalition . . . stretch[ing] from Japan through China, Pakistan, Iran, and Turkey to Western Europe."[99] The reader will smile wryly at the turn of the wheel of history: Mao in 1974 asking for the United States to reclaim the geopolitics of Dulles in 1954!

For their part, the Russians could not believe that their civilized (and Caucasian) Cold War enemy was really going to make up with the Maoist barbarians. In June 1973, Brezhnev warned Richard Nixon not to negotiate a U.S.-China military accord, and a year later he offered the president an alliance against the PRC.[100] One Russian argument spoke to Mao's boasts that China would survive any nuclear war because it had the largest population: "to cause a clash between the U.S. and the U.S.S.R., in order to dominate the world after a nuclear conflict which . . . would reduce America and Europe to ashes while sparing . . . millions of Chinese."[101] Nancy Bernkopf Tucker argues that the huge increase in Soviet military spending that began in the mid-1960s arose in part out of fear of a conventional war with China.[102] Suddenly, in 1972, the prospect of a Sino-American alliance projected the "classic Russian military nightmare—the possibility of a two-front war—made particularly frightful by American and Japanese willingness to arm the Chinese hordes."[103]

In the 1970s, the Soviet Union allocated about 20 to 25 percent of its land units to face China and about one-third of its naval forces to the Pacific.[104] Ronald Reagan was to characterize the seesaw of increasing Soviet military spending while U.S. defense budgets shrank with the remark: "What arms race? We stopped, they raced."[105] The Soviets modernized and extended their nuclear arsenal with large new deployments to Europe.[106] Conventional forces also ballooned: the USSR tank forces introduced new designs and increased in quantity from thirty-four thousand in 1976 to fifty thousand in 1980. The air forces and navy underwent similar upgrades and increases.[107] China took notice but did not follow the Soviet march to bankruptcy. Deng Xiaoping, the architect of post-Mao China, astutely recognized that "one of the reasons why the Soviet economy suffers from paralysis is that the Russians spend too much money on the military."[108] China's defense spending decreased from 10 percent of the GDP in 1971 to less than 2.5 percent through the 1990s.[109] In that sense, Kissinger's power balance was achieved by 1980, but if China had been led by more bellicose men than Deng in the 1970s the outcome might have been different.

The period after Nixon was, of course, a stasis: Gerald R. Ford was an unelected president with enough troubles at home not to risk further China initiatives. In China, the deaths of top leaders and internal political upheavals delayed further foreign policy shifts. But after a new American president broke the last China barrier and recognized the PRC in 1979 (and dropped recognition of Republican China), U.S.-Chinese relations entered a period that we might dare call "normal." Officially this was the name: there would be upsets along the way, more trade fights, and squabbles over the nature of the Taiwan-American relationship, but basically China was no longer a nemesis. Even the last Cold Warrior, President Ronald Reagan, thumping on the campaign trail about the sellout of Taiwan in 1980, would end up a China lover of a sort. And the Middle Kingdom would open up as never before. A new kind of American would flock to China: the tourist. Millions would follow Nixon to visit the Ming Tombs and Tiananmen Square, the Forbidden City, and the Great Wall; most would come home with a T-shirt quoting Mao: "No man is a man until he has climbed the Great Wall."

By the late 1980s, thus, it would seem that the West's oldest dream about China was coming true: China was indeed being remade in the American image. That China was officially run by "Communists" who abrogated basic human rights and crushed a land called Tibet was not relevant; what we saw were Chinese people driving cars, wearing skirts and heels, and eating at McDonald's. Yes, the rapprochement with China was a top-down shape shifting. In terms of human rights, as one writer has put it, "In 1971, when the positive shift in U.S.-China relations occurred and the period of total sanctions during the 1950s and 1960s was drawing to a close, the Chinese government was engaged in its worst period of human rights abuses."[110] But with Mao dead and the "radicals" purged shortly afterward in 1976 there was no figure to whom to attach the old animus. Indeed, China's new, modest leader (who eschewed a personality cult) was the man who said "to get rich is glorious" and "it does not matter if the cat is black or white as long as it catches mice."

If Ronald Reagan could refer, as he did in 1984, to "so-called Communist China," surely Deng's kingdom had changed for the better?[111] Deng's pressing forward with "four modernizations" was unquestionably his attempt to become more like America.[112] *Time* would crown him "Man of the Year" in 1978 for the "tremendous enterprise he has launched to propel the nation into the modern world"[113] and again in 1985 for the success of his economic ventures.[114] Was this not a China to appreciate? Our vision of China had changed so as to be unrecognizable from the 1960s; alas, it proved to be as faulty as ever. The period from 1974 to the mid-'80s did bring about a normal relationship with China, but we must ask whether it was a normalcy that left us unprepared for the problems to come.

These issues seemed paramount when, in the spring of 1989, China was rocked by demonstrations that focused on Beijing's one-hundred-acre epicenter of political struggles, Tiananmen Square. Print and television reporters from all over the world converged to display to an international audience a veritable spectacle of visual drama: student protests rising into the hundreds of thousands of participants, hunger strikes, colorful banners, placards and posters, and a "Goddess of Democracy," signs and symbols that called for a host of reforms ranging from better living conditions at universities to outright democracy. The government leadership, official press, and even the army seemed paralyzed, alternating between decrying the "hooligans and roughnecks" on the streets, praising the brave young people and making a show of reaching out to them, and accusing foreign interests, including the United States, of plotting the extermination of the Communist Party. In the end, the forces of centralized power and repression won the argument, and troops mobilized in the city and cracked down on all forms of dissent. Still today, we have no idea how many people were killed, soldiers included. But thousands were arrested and a pall was cast on Sino-American relations; China's image was tarnished in a way that to some extent lingers to this day. Pictures—photo icons—became instruments for defining the event and the mnemonic device through which "Tiananmen" is recalled in the mind of most of us who lived through that time, and even for another generation today. They include the slogans and signage of the students; calls of "Give Me Liberty or Give Me Death"; the "Goddess of Democracy" erected in the square; and, most of all, a young man, identified as "Wang Weilin," blocking a column of tanks.

But for presidents of the rapprochement era, until June 1989 when George Bush faced the Tiananmen spectacle, China was largely off the table as regards human rights. As Amnesty International put it tersely in a report in the year of Tiananmen: "Since the historic warming of Sino-American relations in 1972, the desire for a strong, stable, and friendly China to check Soviet influence in Asia has overshadowed all other considerations in U.S. policy toward the People's Republic."[115] That is not to say that the situation in China was worse than in, say, 1958 or 1968. The Cultural Revolution was over; so was politically driven famine. In May 1979, Deng proclaimed the Four Cardinal Principles for society to complement (and control) Zhou Enlai's Four Modernizations in the economy: These included "upholding the socialist path," "upholding the people's democratic dictatorship," "upholding the leadership of the Communist party," and "upholding Marxist Leninism Mao Zedong Thought." As developed in the decades to come—and demonstrated in the squashing of various small and large democracy movements and dissident groups—the CCP would remain the state authority, but people could pursue wealth. As several scholars have noted, the intended outcome of such "reforms" was the "establishment of a *system* of law"[116] or "rule by law"[117] rather than "rule of law." Rejected was "bourgeois democracy" and its hypocrisies in favor of the party's control over all political life.

As a consequence, the right of citizens of the PRC to get rich was the only human right from 1972 to spring 1989 that both American presidents and Communist leaders would encourage. But—a big "but"—the politicians did not stand alone in looking the other way on China's human rights failings. Even China experts in the West—academics, for example—made the human rights situation in the country a minor topic of study and commentary.[118] In sum, the accusation of journalist James Mann, that "Carter and his aides gave China virtually a blanket exemption from the human rights policies they so readily applied elsewhere,"[119] is factually correct, but the characterization could be applied to practically all China watchers in the 1970s and 1980s. As Roberta Cohen observed before and after rapprochement, "no systematic or serious effort was made by governments or human rights organizations to call the PRC to account or even to document its abuses. No detailed analysis of China's human rights record appeared."[120] She made this point in 1986; a few years later everyone, rhetorically at least, would discover that China had a human rights problem. Then the long "double standard," or rather "no standard," for China would be painfully exposed, although little reflected upon, in pictures or in policies—unless of course, we take the time, as I hope I have done here, to look at how we looked at China in the past.

Notes

1. G. W. Choudhury, *China in World Affairs: The Foreign Policy of the PRC since 1970* (Boulder, Colo.: Westview Press, 1982), 71–72.
2. Henry Kissinger, *White House Years* (Boston: Little, Brown, 1979), 690.
3. Richard M. Nixon, "The Speeches of Richard M. Nixon," (partial transcript of remarks at Knott's Berry Farm, Long Beach, Calif., October 12, 1960), 555–56.
4. Richard M. Nixon, excerpts from address to American Society of Newspaper Editors, April 20, 1963, in "Excerpts from Address by Nixon Calling for a Stronger U.S. Foreign Policy," *New York Times*, April 21, 1963, A62.
5. James Mann, *About Face: A History of America's Curious Relationship with China, from Nixon to Clinton* (New York: Knopf, 1999), 17; Nancy B. Tucker, ed., *China Confidential: American Diplomats and Sino-American Relations 1945–1996* (New York: Columbia University Press, 2001), 222–23.
6. Richard Nixon, "Asia After Vietnam," *Foreign Affairs* 46, no. 1 (1967): 111–25.
7. *U.S. News & World Report*, September 16, 1968, 48.
8. *New York Times*, May 2, 1968.
9. *New York Times*, April 23, 1968; Banning Garrett, "The 'China Card' and Its Origins" (Ph.D. diss., Brandeis University, 1983).
10. Lyle J. Goldstein, "Do Nascent WMD Arsenals Deter? The Sino-Soviet Crisis of 1969," *Political Science Quarterly* 118, no.1 (2003), 53–79.
11. Richard M. Nixon, *RN: The Memoirs of Richard Nixon* (New York: Grosset and Dunlap, 1978), 284.
12. Henry Kissinger, *A World Restored: Metternich, Castlereagh and the Problems of Peace, 1812-1822* (New York: Houghton Mifflin, 1957).
13. Kissinger, *A World Restored*, 325.
14. Henry A. Kissinger, *Nuclear Weapons and Foreign Policy* (New York: Harper & Brothers/Council on Foreign Relations, 1957), 160.

15. Richard M. Nixon, "U.S. Foreign Policy for the 1970s: A New Strategy for Peace," a report to the Congress, February 18, 1970, 3.

16. Kissinger, *White House Years*, 191.

17. Kissinger, *White House Years*, 1062, 222.

18. *Time*, January 3, 1972, 86.

19. Seymour M. Hersh, *The Price of Power: Kissinger in the White House* (New York: Summit, 1983), 357.

20. Foster Rhea Dulles, *American Foreign Policy toward Communist China, 1949–1969* (New York: Crowell, 1972), 243.

21. League of Women Voters, *Taiwan and American Policy* (New York: Praeger,1971), 13.

22. Congressional Quarterly, *China and the U.S. Far East Policy, 1945–1967* (Washington, D.C.: Congressional Quarterly Service, 1967): 39–41.

23. *New York Times*, April 26, 1970, A1.

24. Kissinger, *White House Years*, 697, 1087.

25. Richard Nixon, *The Memoirs of Richard Nixon*, (New York: Warner, 1979), 2:8.

26. Denise M. Bostdorff, "The Evolution of a Diplomatic Surprise: Richard M. Nixon's Rhetoric on China, 1952–July 15, 1971," *Rhetoric & Public Affairs* 5, no. 1 (2002): 31–56.

27. United States Department of State, "Foreign Relations of the United States: Diplomatic Papers," in vol. 1, *Foreign Relations, 1969–1976; Foundations of Foreign Policy, 1969–1972* (Washington, D.C.: United States Department of State, 2003), www.state.gov/www/about_state/history/vol_xxx/index.html and www.state.gov/r/pa/ho/Foreign Relations of the United States/Nixon/I/.

28. Nixon, *Memoirs*, 2:8.

29. King C. Chen, *China and the Three Worlds* (White Plains, N.Y.: Sharpe, 1979), 159.

30. Kissinger, *White House Years*, 182; Nixon, *Memoirs*, 2:8.

31. United States Department of State, *Foundations of Foreign Policy, 1969–1972.*

32. United States Department of State, *Foundations of Foreign Policy, 1969-1972*; Marvin and Bernard Kalb, *Kissinger* (Boston: Little, Brown, 1974), 232.

33. United States Department of State, *Foundations of Foreign Policy, 1969–1972.*

34. Lawlor A. McNulty, "The Sino-Soviet Border Problems of 1969," *Parameters* 13, no. 3 (1983): 59–67; John W. Garver, *China's Decision for Rapprochement with the United States, 1968-1971* (Boulder, Colo.: Westview, 1982), 69.

35. Alfred D. Low, *The Sino-Soviet Confrontation Since Mao Zedong* (New York: Columbia University Press, 1981), 41.

36. United States Department of State, *Foundations of Foreign Policy, 1969–1972.*

37. Kissinger, *White House Years*, 188.

38. Cited in John Maxwell Hamilton, *Edgar Snow: A Biography* (Bloomington: Indiana University Press, 1988), 265.

39. Jian Chen and David L. Wilson, "'All under the Heaven Is Great Chaos': Beijing, the Sino-Soviet Border Clashes, and the Turn toward Sino-American Rapprochement, 1968–69," *Cold War International History Project Bulletin* 11 (1998): 155–75, www.cwihp.si.edu/cwihplib.nsf/1 . . . 7f2a28525677c00686a69, 8.

40. "Conversation between Mao Zedong and E. F. Hill," in "'All under Heaven Is Great Chaos': Beijing, the Sino-Soviet Border Clashes, and the Turn toward Sino-American Rapprochement, 1968–69," *Cold War International History Project*, www.cwihp.si.edu/cwihplib.nsf/1 . . . 7f2a28525677c00686a69, 8..

41. Xiong Xianghui, *Wo De Qingbao Yu Waijiao Shengya* [My intelligence and diplomatic career], trans. Shannon Wang (Beijing: Zhongyang Dangxiao Chubanshe, n.d.), 184–86.

42. Yafeng Xia, "From Antagonists to Adversaries: United States-China Talks During the Cold War, 1949–1972" (Ph.D. diss., University of Maryland, 2003), 190.

43. Yafeng Xia, "From Antagonists to Adversaries," 245.

44. Kissinger, *White House Years*, 687.

45. Harold C. Hinton, *China's Turbulent Quest: An Analysis of China's Foreign Relations Since 1949* (Bloomington, Indiana University Press, 1972), 287, 290.

46. Joseph Camillieri, *Chinese Foreign Policy: The Maoist Era and Its Aftermath* (Seattle: University of Washington Press, 1989), 173.

47. Nien Cheng, *Life and Death in Shanghai* (New York: Grove Press, 1986), 306.

48. Dino A. Brugioni, *Photo Fakery: The History and Techniques of Photographic Deception and Manipulation* (Dulles, Va.: Brassey's, 1999); Alain Jaubert, *Making People Disappear: An Amazing Chronicle of Photographic Deception* (Washington, D.C.: Pergamon-Brassey's, (1989[1986]).

49. Kissinger, *White House Years*, 698–99; Edgar Snow, "China Will Talk from a Position of Strength," *Life*, July 30, 1971, 24.

50. Yawei Liu, "The United States according to Mao Zedong: Chinese-American Relations, 1893–1976" (Ph.D. diss., Emory University, 1996), 467–68.

51. Tad Szulc, "Zhou Said to Back Parley with U.S.," *New York Times*, June 18, 1970, A9.

52. Kissinger, *White House Years*, 192.

53. United States Department of State, *Foundations of Foreign Policy, 1969–1972.*

54. CQ, *China and the U.S. Far East Policy, 1945–1967* (Washington, D.C.: Congressional Quarterly Service, 1967): 13–15.

55. CQ, *China and the U.S. Far East Policy, 1945–1967*, 103.

56. Ruth Eckstein, "Ping-Pong Diplomacy: A View from behind the Scenes," *Journal of American-East Asian Relations* 2, no. 3 (1993): 327–42.

57. Kissinger, *White House Years*, 712.

58. Yawei Liu, "The United States according to Mao Zedong," 472.

59. Richard Nixon, *Memoirs*, 2:17.

60. United States Department of State, *Foundations of Foreign Policy, 1969–1972.*

61. Richard Madsen, *China and the American Dream: A Moral Inquiry* (Berkeley: University of California Press, 1995), 65.

62. United States Department of State, *Foundations of Foreign Policy, 1969–1972.*

63. United States Department of State, *Foundations of Foreign Policy, 1969–1972.*

64. Kissinger, *White House Years*, 766–67; United States Department of State, *Foundations of Foreign Policy, 1969–1972.*

65. See: Hersh, *The Price of Power*, 378; Nancy B. Tucker, ed., *China Confidential: American Diplomats and Sino-American Relations 1945–1996* (New York: Columbia University Press, 2001), 264.

66. Russell Maud, "The Why of Nixon's Visits to China," *Far East Report,* July 1–8, 1972, 7.

67. William R. Keech and Donald R. Matthews, *The Party's Choice* (Washington, D.C.: Brookings Institution, 1976), 89.

68. Madsen, *China and the American Dream,* 69.

69. Madsen, *China and the American Dream,* 73.

70. Richard Nixon, *RN: The Memoirs of Richard Nixon* (New York: Grosset and Dunlap, 1978), 557–58, 560; H. R. Haldeman, *The Haldeman Diaries* (New York: G. P. Putnam's, 1994), 411–13.

71. Madsen, *China and the American Dream,* 73.

72. Richard Nixon, *Public Papers of the President of the United States: Richard Nixon 1972.* (Washington, D.C.: Government Printing Office), 379.

73. Nixon, *RN,* 562

74. Cited in Zhisui Li, *The Private Life of Chairman Mao,* trans. Tai Hung-Chao (New York: Random House, 1994), 564.

75. Cited in Li, *The Private Life of Chairman Mao,* 563–64.

76. Cited in Li, *The Private Life of Chairman Mao,* 565.

77. Kissinger, *White House Years,* 1016.

78. Walter Isaacson, *Kissinger: A Biography* (New York: Simon & Schuster, 1992), 404–5; William Bundy, *A Tangled Web: The Making of Foreign Policy in the Nixon Presidency* (New York: Hill and Wang, 1992), 305–6; Kissinger, *White House Years,* 1074–86, 1092, 1094–95.

79. Henry Kissinger, *White House Years* (Boston: Little, Brown, 1979), 164–65.

80. G. W. Choudhury, *China in World Affairs: The Foreign Policy of the PRC Since 1970* (Boulder, Colo.: Westview, 1982), 71–72; Kissinger, *White House Years,* 164–65.

81. Liqun Deng, *China's Diplomacy Today* (Beijing: Chinese Social Science Publishing House, 1987), 222.

82. USDA and U.S. Economic Research Service, "Wheat Situation," (November 1972), 30.

83. Martin Jurek, "Nixon's Boeing Move," *Far Eastern Economic Review,* July 15, 1972, 31.

84. Doak A. Barnett, *China and the Major Powers in East Asia* (Washington, D.C.: Brookings Institution, 1977), 224.

85. Stanley Karnow, "Changing (Mis)Conceptions of China," in *The Image of China,* ed. Benson Grayson (New York: Ungar, 1979), 284–85.

86. Stanley D. Bachrack, "The Death Rattle of the China Lobby," *Washington Post,* February 2, 1979.

87. Cited in John Maxwell Hamilton, *Edgar Snow: A Biography* (Bloomington: Indiana University Press, 1988), 277.

88. John K. Fairbank, *China Perceived: Images and Policies in Chinese-American Relations* (New York: Knopf, 1974), xvi.

89. Warren Cohen, "American Perceptions of China" in *Dragon and Eagle: U.S.-China Relations: Past and Future,* ed. Michel Oksenberg and Robert B. Oxnam (New York: Basic Books, 1978), 54–86, 83.

90. Kissinger, *White House Years,* 1062.

91. Kissinger, *White House Years*, 222.

92. G. W. Choudhury, "China Revisited," *Orbis* 10, no. 3 (1976): 606.

93. J. K. Fairbank, *The United States and China*, 4th ed. (Cambridge, Mass.: Harvard University Press, 1983), 320.

94. Stephen E. Ambrose, *Nixon*, vol. 2 (New York: Simon & Schuster, 1989), 440–41.

95. "Like Nixon to China" was to become a catchphrase and a ritual allusion for someone who is able to accomplish a task because of his personal qualities and past. John Kerry, on the campaign trail, stated, "Remember how only Nixon could go to China? . . . only John Edwards and John Kerry can get tort reform done in America, in a way that's fair." See: www.nixonlibrary.org/nrc/041018akerry.shtml.

96. David D. Perlmutter, *Photojournalism and Foreign Policy: Framing Icons of Outrage in International Crises* (Westport, Conn.: Greenwood, 1998).

97. Henry Kissinger, *Diplomacy* (New York: Simon & Schuster, 1994), 461.

98. Henry Kissinger, *Years of Upheaval* (Boston: Little, Brown, 1982), 70.

99. Henry Kissinger, *Years of Upheaval*, 55, 67.

100. William Hyland, *Mortal Rivals* (New York: Random House, 1987), 11, 60, 63–64; Richard Nixon, *RN: The Memoirs of Richard Nixon* (New York: Grosset & Dunlap, 1978), 1030; John Newhouse, *Cold Dawn* (New York: Holt, Rinehart and Winston, 1973), 188–89.

101. Banning Garrett and Bonnie Glaser, *War and Peace* (Berkeley: University of California Press, 1984), 28.

102. Nancy B. Tucker, "China as a Factor in the Collapse of the Soviet Empire," *Political Science Quarterly* 110 (1995): 501–18.

103. Nancy B. Tucker, "China as a Factor"

104. Chi-Wu Wang, "Military Preparedness and Security Needs: Perceptions from the Republic of China on Taiwan," *Asian Survey* 21, no. 6 (1981): 651.

105. Quoted in John Spanier, *American Foreign Policy Since World War II*, 11th ed. (Washington, D.C.: Congressional Quarterly, 1988): 274.

106. International Institute for Strategic Studies, "The Military Balance, 1980" (London: IISS, 1980), 3–4.

107. International Institute for Strategic Studies, "The Military Balance, 1980," 5.

108. Thomas Bernstein, "Domestic Politics," in *China Briefing, 1984*, ed. Steven Goldstein (Boulder, Colo.: Westview, 1985), 9; Samuel S. Kim, "China and the Third World: In Search of a Peace and Development Line," in *China and the World*, ed. Samuel S. Kim (Boulder, Colo.: Westview Press, 1989), 169.

109. James H. Nolt, "Liberalizing Asia," *World Policy Journal* 16, no. 2 (Summer 1999): 94–110.

110. James D. Seymour, "Sanctions or Subdued Relations: The International Response to the 1989 Massacre," *Fletcher Forum of World Affairs* 14, no. 1 (1990): 55–61.

111. Mann, *About Face*, 146–47.

112. Andrew J. Nathan, *China's Crisis: Dilemmas of Reform and Prospects for Democracy* (New York: Columbia University Press, 1990).

113. "Time's 'Man of the Year for 1978,' Teng Hsiao-P'ing: Visionary of a New China," *Time*, January 1, 1979, 12–39.

114. "Time's 'Man of the Year for 1978,'" 24–50, 59–61.

115. Quoted by Marie Gottschalk, "The Failure of American Policy," *World Policy Journal* 6, no. 4 (1989): 667–85.

116. Minxin Pei, "Creeping Democratization in China," *Journal of Democracy* 6, no. 4 (1995): 65–79.

117. James V. Feinerman, "The Rule of Law . . . with Chinese Socialist Characteristics," *Current History* 96, no. 611 (1997): 278–81.

118. See: Merle Goldman, "The Persecution of China's Intellectuals: Why Didn't Their Western Colleagues Speak Out?" *Radcliffe Quarterly* 67, no. 3 (1981): 12–14; see also: Roberta Cohen, "People's Republic of China: The Human Rights Exception," *Human Rights Quarterly* 9, no. 3 (1987): 467–70.

119. Mann, *About Face*, 100.

120. Roberta Cohen, "People's Republic of China."

Bibliography

Acheson, Dean. The Korean War. New York: W. W. Norton, 1971.

———. *Present at the Creation: My Years in the State Department.* New York: W. W. Norton, 1969.

Aldridge, A. O. *The Dragon and the Eagle: The Presence of China in the American Enlightenment.* Detroit: Wayne State University Press, 1993.

Alexander, Joseph H., Don Horan, and Norman C. Stahl. *The Battle History of the U.S. Marines: A Fellowship of Valor.* New York: Harper Perennial, 1999.

Alsop, Joseph. "The Feud between Stilwell and Chiang." *Saturday Evening Post,* January 7, 1950, 16.

Ambrose, Stephen E. *Eisenhower, the President.* Vol. 2. New York: Simon & Schuster, 1984.

———. *Nixon,* Vol. 2. New York: Simon & Schuster, 1989.

American Foreign Policy. *Current Documents, 1956-1967.* Washington, D.C.: Government Printing Office, 1958.

Amsden, A. "Taiwan's Economic History." *Modern China* 5 (1979): 341–80.

Anderson, D. L. *Imperialism and Idealism: American Diplomats in China, 1861–1898.* Bloomington: Indiana University Press, 1985.

Appleman, Roy C. *Ridgway Duels for Korea.* College Station: Texas A&M University Press, 1990.

Appleman, Roy E. *1950: United States Army in the Korean War: South to the Naktong, North to the Yalu.* Washington, D.C.: U.S. Army, Center of Military History, 1992.

Arnold, William A. *Back When It All Began.* New York: Vantage, 1975.

Bachrack, Stanley D. *The Committee of One Million: China Lobby Politics, 1953–1971.* New York: Columbia University Press, 1976.

———. "The Death Rattle of the China Lobby." *Washington Post,* February 2, 1979.

Ball, Moya. *Vietnam-On-the-Potomac.* Westport, Conn.: Praeger, 1992.

Bank of Taiwan. *Three States of Taiwan's Cession.* Taipei: Ko Tai San Chi, 1959.

Bannister, Judith. "Population Policy and Trends in China, 1978–1983." *China Quarterly* 100, (December 1984): 717–42.

Barnett, A. Doak. *China and the Major Powers in East Asia.* Washington, D.C.: Brookings Institution, 1977.

———. "The 1958 Quemoy Crisis: The Sino-Soviet Dimension." *Problems of Communism* 25 (July/August 1976): 38.

Barnouw, Erik. *Documentary: A History of the Non-Fiction Film.* New York: Oxford University Press, 1983.

Barrett, David D. *Dixie Mission: The United States Army Observer Group in Yanan, 1944.* Berkeley, Calif.: Center for Chinese Studies, 1970.

Baughman, J. L. *Henry R. Luce and the Rise of the American News Media.* Boston: Twayne, 1987.

Becker, Jasper. *Hungry Ghosts: China's Secret Famine.* New York: Henry Holt, 1998.

Bergerud, Eric M. *The Dynamics of Defeat: The Vietnam War in Hau Nghia Province.* Boulder, Colo.: Westview Press, 1991.

Bernstein, Thomas P. "Domestic Politics." in *China Briefing, 1984,* edited by Steven Goldstein. Boulder, Colo.: Westview Press, 1985.

———. "Stalinism, Famine, and Chinese Peasants: Grain Procurements During the Great Leap Forward." *Theory and Society* 13, no. 3 (1984): 339–77.

Best, Gary Dean. *Pride, Prejudice, and Politics: Roosevelt Versus Recovery, 1933–1938.* Westport, Conn.: Praeger, 1993.

Bienstock, Gregory. *The Struggle for the Pacific. Port Washington.* New York: Kennikat Press, 1937.

Blair, Clay. *The Forgotten War: America in Korea, 1950–53.* New York: Time Books, 1988.

Blum, Robert M. *Drawing the Line: The Origins of the American Containment Policy in East Asia.* New York: W. W. Norton, 1982.

Borg, Dorothy, ed. *Uncertain Years: Chinese-American Relations. 1947–1950.* New York: Columbia University Press, 1970.

Bostdorff, Denise M. "The Evolution of a Diplomatic Surprise: Richard M. Nixon's Rhetoric on China. 1952- July 15, 1971." *Rhetoric & Public Affairs* 5, no. 1 (2002): 31–56.

Bowles, Chester. "The China Problem Reconsidered." *Foreign Affairs* 38, no. 3 (1960): 476–86.

Bradley, Omar N. *A General's Life: An Autobiography.* New York: Simon & Schuster, 1983.

Brady, Anne-Marie. *Making the Foreign Serve China: Managing Foreigners in the People's Republic.* Lanham, Md.: Rowman & Littlefield, 2003.

Brugioni, Dino A. *Photo Fakery: The History and Techniques of Photographic Deception and Manipulation.* Dulles, Va.: Brassey's, 1999.

Buckley, William F., Jr. *Execution Eve and Other Contemporary Ballads.* New York: G. P. Putnam's Sons, 1975.

Bundy, William. *A Tangled Web: The Making of Foreign Policy in the Nixon Presidency.* New York: Hill and Wang, 1992.

Burns, James McGregor. *John Kennedy: A Political Profile.* New York: Avon, 1959.

Camillieri, Joseph. *Chinese Foreign Policy: The Maoist Era and Its Aftermath.* Seattle: University of Washington Press, 1989.

Capra, Frank. *The Name above the Title: An Autobiography.* New York: Macmillan, 1971.

Ch'i, Hsi-Sheng. "The Military Dimension, 1942–1945." In *China's Bitter Victory: The War with Japan, 1937–1945,* edited by James C. Hsiung and Steven I. Levine. Armonk, 157–84. N.Y.: M. E. Sharpe, 1992.

Chang, Gordon H. *Friends and Enemies: The United States. China and the Soviet Union, 1948–72.* Stanford, Calif.: Stanford University Press, 1990.

———. "JFK, China and the Bomb." *Journal of American History* 74, no. 4 (1999): 1287–310.

Chang, Hsien-Wen, ed. *K'ang-Jih Chan-Cheng Ti Cheng-Mien Chan-Ch'ang 1959.* Translated by Shannon Wang. N.P.: Honan Jen-Min Ch'u-Pan-She, 1987.x

Chay, Jongsuk. *Unequal Partners in Peace and War: The Republic of Korea and the United States, 1948–1953.*Westport, Conn.: Praeger, 2002.

Chen, Jian. "China's Road to the Korean War: A Critical Study of the Origins of Sino-American Confrontation, 1949–1950." Ph.D. diss., Southern Illinois University at Carbondale, 1990.

———. *China's Road to the Korean War: The Making of the Sino-American Confrontation.* New York: Columbia University Press, 1990.

———. "The Sino-Soviet Alliance and China's Entry into the Korean War." Working paper, Cold War History Project, Woodrow Wilson International Center For Scholars.

Chen, Jian, and David L. Wilson. "'All Under the Heaven is Great Chaos': Beijing, the Sino-Soviet Border Clashes, and the Turn toward Sino-American Rapprochement, 1968–69." *Cold War International History Project Bulletin* 11 (1998): 155–75. www.cwihp.si.edu/cwihplib.nsf/1...7f2a28525677c00686a69, 8.

Chen, King C. *China and the Three Worlds.* Armonk, N.Y.: M. E. Sharpe, 1979.

Chen, Mingtong. "Paixi Zhengzhi Yu Chen Yi Zhi Tai Lun" [A discussion of political factions and Chen Yi's rule of Taiwan]. Translated by Shannon Wang. In *Taiwan Guangfu Chuqi Lishi,* edited by Jeh-Han Lai, 257–59. Nangang: Zhongyang Yanjiuyuan Zhongshan Renwen Shehui Kexue Yanjiusuo, 1993.

Cheng, Nien. *Life and Death in Shanghai.* New York: Grove Press, 1986.

Chien, Hou Tsung. "U.S. Policy toward China, 1912 to Present." Ph.D. diss., Northern Arizona University, 1985.

China and U.S. Far East 1945–1967. Washington, D.C.: Congressional Quarterly Service, 1967.

Chinese Academy of Military Sciences, ed. *The PLA History of the War to Resist America and Aid Korea.* Beijing: Junshi Kexue Chubanshe, 1988.

Chinese Intervention. U.S. Army. Korean War, Commenoration Brochures. November 3, 1950–January 24, 1951. www.army.mil/cmh-pg/brochures/kw-chinter/chinter.htm.

Ching, J., and W. G. Oxtoby, eds. *Discovering China: European Interpretations in the Enlightenment.* Rochester, N.Y.: University of Rochester Press, 1992.

Chiu, Hungdah. ed. *China and the Question of Taiwan: Document and Analysis.* New York: Praeger, 1973.

———, *China and the Taiwan Issue.* New York: Praeger, 1979.

Choudhury, G. W. *China in World Affairs: The Foreign Policy of the PRC Since 1970.* Boulder, Colo.: Westview Press, 1982.

Christensen, Thomas J. "Domestic Mobilization and International Conflict: Sino-American Relations in the 1950s." Ph.D. diss., Columbia University, 1993.

Churchill, Winston S. *Closing the Ring.* Boston: Houghton Mifflin, 1951.

———. *The Second World War.* Vol. 4, *The Hinge of Fate.* London: Cassell, 1951.

Cohen, E. A., and J. Gooch. *Military Misfortunes: The Anatomy of Failure in War.* New York: Vintage, 1991.

Cohen, Paul A. "Cultural China: Some Definitional Issues." *Philosophy East & West* 43 (1993): 557.

Cohen, Roberta. "People's Republic of China: The Human Rights Exception." *Human Rights Quarterly* 9, no. 4: 447–549.

Cohen, Warren I. "Acheson, His Advisers, and China, 1949–1950." In *Uncertain Years: Chinese-American Relations. 1947-1950,* edited by Dorothy Borg, 49–51. New York: Columbia University Press, 1980.

———. "American Perceptions of China." In *Dragon and Eagle: U.S.-China Relations: Past and Future,* edited by Michel Oksenberg and Robert B. Oxnam, 54–86. New York: Basic Books, 1978.

———. *America's Response to China: A History of Sino-American Relations.* New York: Columbia University Press, 1990.

———. "The China Lobby." In *Encyclopedia of American Foreign Policy: Studies of the Principal Movements and Ideas,* edited by A. Deconde, 104–10. New York: Scribner, 1978.

———. *The Chinese Connection.* New York: Columbia University Press, 1978.

———. "Research Note: Conversations with Chinese Friends: Zhou Enlai's Associates Reflect on Chinese-American Relations in the 1940s and the Korean War." *Diplomatic History* 11, no. 3 (1987): 285.

Cold War International History Project. www.cwihp.si.edu/cwihplib.nsf/1...7f2a28525677c00686a69, 8.

Colombo, Claudius. "Chinese Communist Perceptions of the Foreign Policy of John F. Kennedy, 1961–1963." Ph.D. diss., New York University, 1982.

Condit, Doris M. *History of the Office of the Secretary of Defense: The Test of War, 1953.* Washington D.C.: Office of the Secretary of Defense, 1988.

Congressional Quarterly. *China and the U.S. Far East Policy, 1945–1967.* Washington, D.C.: Congressional Quarterly Service, 1967.

Craft, Stephen G. "Peacemakers in China: American Missionaries and the Sino-Japanese War, 1937–1941." *Journal of Church and State,* 41 (1999): 575–91.

Cumings, Bruce. *Korea's Place in the Sun: A Modern History.* New York: W. W. Norton, 1997.

Dallek, Robert. *Flawed Giant: Lyndon Johnson and His Times.* New York: Oxford University Press, 1998.

———. "JFK's Second Term." *Atlantic Monthly,* June 2003, 58–66.

———. *An Unfinished Life: John F. Kennedy, 1917–1963.* New York: Oxford University Press, 2003.

Defourneaux, Rene J. *The Winking Fox.* Indianapolis: Indiana Creative Arts, 1977.

Deng, Liqun. *China's Diplomacy Today.* Beijing: Chinese Social Science Publishing House, 1987.

Diamond, Robert A., ed. *China and U.S. Foreign Policy.* Washington, D.C.: Congressional Quarterly, 1971.

Doherty, Thomas. *Projections of War: Hollywood, American Culture and World War II.* New York: Columbia University Press, 1993.

Donovan, Robert J. *Eisenhower: The Inside Story.* New York: Harper, 1956.

Dulles, Foster Rhea. *American Policy toward Communist China.* New York: Thomas Y. Crowell, 1979.

———. *China and America: The Story of Their Relation Since 1784.* Washington, D.C.: Kennikat Press, 1967.

Durdin, Tillman. "Zhou Asks For U.S. Talks on Formosa Crisis; Washington Sets Terms." *New York Times,* April 24, 1955, 1.

Eckstein, Ruth. "Ping-Pong Diplomacy: A View from behind the Scenes." *Journal of American-East Asian Relations* 2, no. 3 (1993): 327–42.

Eisenhower, Dwight D. *The White House Years.* Vol. 1, *Mandate for Change.* Garden City, N.Y.: Doubleday, 1963.

———. *The White House Years.* Vol. 2, *Waging Peace, 1956–1961.* New York: Doubleday, 1965.

Eldridge, Fred. *Wrath in Burma: The Uncensored Story of General Stilwell and International Maneuvers in the Far East.* New York: Doubleday, 1946.

Esherick, Joseph W., ed. *Lost Chance in China: The World War II Despatches of John S. Service.* New York: Random House, 1974.

Etzold, Thomas, and John Lewis Gaddis, eds. "Conversations between General of the Army MacArthur and Mr. George F. Kennan." In *Containment: Documents On American Policy and Strategy, 1945–1950.* New York: Columbia University Press, 1978.

Fairbank, John K. *China Perceived: Images and Policies in Chinese-American Relations.* New York: Knopf, 1974.

———. "Our Inherited China Policy." In *The United States and China,* 307–35. Cambridge, Mass.: Harvard University Press, 1983.

———. *The United States and China.* 4th ed. Cambridge, Mass.: Harvard University Press, 1983.

Fautua, David T. "The 'Long Pull' Army. Nsc-68, the Korean War, and the Creation of the Cold War U.S. Army," *Journal of Military History* 61, no. 1 (January 1997): 95.

Fehrenbach, T. R. *This Kind of War: A Study in Unpreparedness.* New York: Macmillan, 1963.

Feinerman, James V. "The Rule of Law . . . with Chinese Socialist Characteristics." *Current History* 96, no. 611 (1997): 278–81.

Feis, Herbert. *The China Tangle: The American Effort in China from Pearl Harbor to the Marshall Mission.* Princeton, N.J.: Princeton University Press, 1953.

Fenby, Jonathan. *Chiang Kai-Shek: China's Generalissimo and the Nation He Lost.* New York: Carroll & Graf, 2004.

Fetzer, James. "Clinging to Containment: China Policy." In *Kennedy's Quest For Victory: American Foreign Policy, 1961–1963,* edited by Thomas G. Paterson, 178–97. New York: Oxford University Press, 1989.

Foot, Rosemary. *The Practice of Power: U.S. Relations with China Since 1949.* Oxford: Clarendon Press, 1995.

Freedman, Lawrence. *Kennedy's Wars: Berlin, Cuba, Laos, and Vietnam.* New York: Oxford University Press, 2000.

Freeman, Charles, Jr. "The Process of Rapprochement: Achievements and Problems." In *Sino-American Normalization and Its Policy Implications*, edited by Gene T. Hsiao and Michael Witunsky, 2. Westport, Conn.: Praeger, 1983.

Friedman, Donald J. *The Road from Isolation: The Campaign of the American Committee for Non-Participation in Japanese Aggression, 1938–1941.* Cambridge, Mass.: Harvard University Press, 1968.

Fu, Jen-Kun. *Taiwan and the Geopolitics of the Asian-American Dilemma.* Westport, Conn.: Praeger, 1992.

Gaddis, John L. *Strategies of Containment: A Critical Appraisal of Postwar American National Security Policy.* New York: Oxford University Press, 1982.

Garrett, Banning. "The 'China Card' and Its Origins." Ph.D. diss., Brandeis University, 1983.

Garrett, Banning, and Bonnie Glaser. *War and Peace.* Berkeley: University of California Press, 1984.

Garver, John W. *China's Decision For Rapprochement With the United States, 1968–1971.* Boulder, Colo.: Westview Press, 1982.

———. "Polemics, Paradigms, Responsibility, and the Origins of the U.S.-PRC Confrontation in the 1950s." *Journal of American-East Asian Relations* 3, no. 1 (1994): 1–34.

Gavin, James M. *War and Peace in the Space Age.* London: Hutchinson, 1959.

General of the Army Omar N. Bradley, Chairman of the Joint Chiefs of Staff Substance of statements, Wake Island Conference, October 15, 1950. Compiled from notes kept by the conferees from Washington. Papers of George M. Elsey, 2. www.trumanlibrary.org/wake/wi198_2.htm.

George, Alexander L., David K. Hall, and William E. Simons. *The Limits of Coercive Diplomacy.* Boston: Little, Brown, 1971.

George, Alexander L., and Richard Smoke. *Deterrence in American Foreign Policy: Theory and Practice.* New York: Columbia University Press, 1974.

Gillette, Michael L. "Ellsworth Bunker Oral History Interview I." Transcript, December 9, 1980. Lyndon Baines Johnson Library. 128.83.78.10/Johnson/Archives.Hom/Oral history.Hom/Bunker%2de/Bunker1.pdf.

———. "Lawrence F. O'Brien Oral History Interview," Transcript 6. February 11, 1986. Lyndon Baines Johnson Library. 128.83.78.10/Johnson/Archives.Hom/Oralhistory.Hom/ObrienI/Obrien06.pdf.

Gittinger, Ted. "Peter Vraestrup Oral History Interview I." Accession number 99-03, March 1, 1982. Lyndon Baines Johnson Library. 128.83.78.10/Johnson/Archives.Hom/Oralhistory.Hom/Braestrup/Braestrup.pdf.

Gittings, John. *The World and China, 1922–1972.* New York: Harper & Row, 1974.

Godwin, Paul. "Change and Continuity in Chinese Military Doctrine: 1949–1999." In *Chinese Warfighting: The PLA Experience Since 1949*, edited by David M. Finkelstein, Michael A. McDevitt, and Mark A. Ryan, 23–55. Armonk, N.Y.: M. E. Sharpe, 2001.

Gold, T. *State and Society in the Taiwan Miracle.* Armonk, N.Y.: M. E. Sharpe, 1986.

Goldman, Merle. "The Persecution of China's Intellectuals: Why Didn't Their Western Colleagues Speak Out?" *Radcliffe Quarterly* 67, no. 3 (1981): 12–14.

Goldstein, Lyle J. "Do Nascent WMD Arsenals Deter? The Sino-Soviet Crisis of 1969." *Political Science Quarterly* 118, no. 1 (2003): 53–79.

Goncharov, Sergei N., John W. Lewis, and Xue Litai. *Uncertain Partners: Stalin, Mao, and the Korea War.* Stanford, Calif.: Stanford University Press, 1993.

Gottschalk, Marie. "The Failure of American Policy." *World Policy Journal* 6, no. 4 (1989): 667–85.

Gray, J. *Rebellions and Revolutions: China from the 1800s to the 1980s.* New York: Oxford University Press, 1990.

Griffith, Thomas. *Harry and Teddy: The Turbulent Friendship of Press Lord Henry R. Luce and His Favorite Reporter, Theodore H. White.* New York: Random House, 1995.

Haldeman, H. R. *The Haldeman Diaries.* New York: G. P. Putnam's Sons, 1994.

Hamilton, J. M. *Edgar Snow: A Biography.* Bloomington: Indiana University Press, 1988.

Hamilton, Nigel. *JFK: Reckless Youth.* New York: Random House, 1992.

Han, Suying. *Birdless Summer.* New York: Putnam, 1968.

Hao, Yufan. "Solving the Dilemma in China Policy: 1978–1979. A Case Study of Normalization of U.S.-China Relations and the Taiwan Relations Act" Ph.D. diss., Johns Hopkins University, 1990.

Hao, Yufan, and Zhihai Zhai. "China's Decision to Enter the Korean War: History Revisited." *China Quarterly* 121 (1990): 100.

Hardin, T. H. "American Press and Public Opinion in the First Sino-Japanese War." *Journalism Quarterly* 50 (January 1973): 54–59.

Heinrichs, Waldo. "Eisenhower and the Sino-American Confrontation." In *The Great Powers in East Asia, 1953–1960,* edited by Warren I. Cohen and Akira Iriye, 86–103. New York: Columbia University Press, 1990.

Hersh, Seymour M. *The Price of Power: Kissinger in the White House.* New York: Summit, 1983.

Hilsman, Roger. *To Move a Nation: The Politics of Foreign Policy in the Administration of John F. Kennedy.* Garden City, N.Y.: Doubleday, 1967.

Hinton, Harold C. *China's Turbulent Quest: An Analysis of China's Foreign Relations Since 1949.* Bloomington: Indiana University Press, 1972.

Hoffman, Jon T. *Chesty: The Story of Lt. General Lewis B. Puller.* New York: Random House, 2001.

Holsti, O. R. "The Belief System and National Images: A Case Study." *Journal of Conflict Resolution* 6 (1962): 244–52.

Hsieh, Alice Langley. *Communist China's Strategy in the Nuclear Era.* Englewood Cliffs, N.J.: Prentice Hall, 1962.

———. "The Sino-Soviet Nuclear Dialogue, 1963." In *Sino-Soviet Military Relations,* edited by Raymond L. Garthoff. New York: Praeger, 1966.

Hsu, I. "Chinese Colonization of Taiwan." Ph.D. diss., University of Chicago, 1975.

———. *The Rise of Modern China.* New York: Oxford University Press, 1983.

Hull, Cordell. *The Memoirs of Cordell Hull.* Vol. 2. New York: Macmillan, 1948.

Hummel, A. ed. *Eminent Chinese of the Ch'ing Period.* Washington, D.C.: Government Printing Office, 1943.

Hunt, Michael H. *The Making of a Special Relationship: The United States and China to 1914.* New York: Columbia University Press, 1983.

Hyland, William. *Mortal Rivals.* New York: Random House, 1987.

Important Documents Concerning the Question of Taiwan. Beijing: Foreign Language Press, 1960.

———. Beijing: Foreign Language Press, 1955.

International Institute for Strategic Studies. "The Military Balance, 1980." London: IISS, 1980.

Isaacs, H. R. *Scratches on Our Minds: American Images of China and India.* New York: J. Day, 1958.

Isaacson, Walter. *Kissinger: A Biography.* New York: Simon & Schuster, 1992.

Israel, J. "Mao's Mr. America: Edgar Snow's Images of China." *Pacific Historical Review* 47, no. 1 (1978): 107–22.

Jacoby, Neil. *U.S. Aid to Taiwan.* New York: Praeger, 1966.

Jan, G. P. "Public Opinion and American Policy toward the Sino-Japanese War, 1937–1945." *Asian Profile* 7 (April 1979): 319–34.

Jaubert, Alain. *Making People Disappear: An Amazing Chronicle of Photographic Deception.* Washington, D.C.: Pergamon-Brassey's, 1989[1986].

Johnson, Alexis. *The Right Hand of Power.* New York: Free Press, 1984.

Johnston, Alastair Iain. "Cultural Realism and Strategy in Maoist China." In *The Culture of National Security: Norms and Identity in World Politics,* edited by Peter J. Katzenstein, 216–70. New York: Columbia University Press, 1996.

Jurek, Martin. "Nixon's Boeing Move," *Far Eastern Economic Review,* July 15, 1972, 31.

Kahn, E. J., Jr. *The China Hands: America's Foreign Service Officers and What Befell Them.* New York: Viking Press, 1975.

Kalb, Marvin, and Bernard Kalb. *Kissinger.* Boston: Little, Brown, 1974.

Kalicki, J. H. *The Pattern of Sino-American Crises: Political-Military Interactions in the 1950s.* London: Cambridge University Press, 1975.

Kamber, Victor. *Poison Politics: Are Negative Campaigns Destroying Democracy?* New York: Plenum Press, 1997.

Kapitsa, M. S. *On Different Latitudes—Diplomat's Notes.* Moscow: Kniga i biznes, 1996.

Karnow, Stanley. "Changing (Mis)Conceptions of China." In *The Image of China,* edited by Benson Grayson, 284–85. New York: Ungar, 1979.

———. *Vietnam: A History.* New York: Penguin, 1983.

Karol, K. S. *The Second Chinese Revolution.* Translated by Mervyn Jones. New York: Hill and Wang, 1973.

Kearns, Doris. *Lyndon Johnson and the American Dream.* New York: Harper & Row, 1976.

Keech, William R., and Donald R. Matthews. *The Party's Choice.* Washington, D.C.: Brookings Institution, 1976.

Keeley, Joseph C. *The China Lobby Man: The Story of Alfred Kohlberg.* New Rochelle, N.Y.: Arlington House, 1969.

Kehr, Harvey. *The Amerasia Spy Case: Prelude to McCarthyism.* Chapel Hill: University of North Carolina Press, 1996.

Kelman, Hebert C. "The Social-Psychological Approaches to the Study of International Relations: The Questions of Relevance." In *International Behavior: A Social-Psychological Analysis*, edited by Hebert C. Kelman, 565–607. New York: Holt, Rinehart and Winston, 1965.

Kennedy, John F. "A Democrat Looks at Foreign Policy," *Foreign Affairs* 36. no. 1 (1957): 50.

Kerpen, Karen Shaw. "Voices in a Silence: American Organizations That Worked for Diplomatic Recognition of the People's Republic of China by the United States, 1945–1979." Ph.D. diss., New York University, 1981, 147–48.

Kerr, George H. *Formosa: Licensed Revolution & the House Rule Movement, 1895–1945*. Honolulu: University Press of Hawaii, 1974.

Kessler, H. *K'ang-Hsi and the Consolidation of Ch'ing Rule, 1661–1684*. Chicago: University of Chicago Press, 1976.

Khrushchev, Nikita S. *Khrushchev Remembers*. Translated edited by Edward Crankshaw and Strobe Talbbott. Boston: Little, Brown, 1970.

———. *Khrushchev Remembers: The Last Testament*. London: Deutsch, 1974.

Kim, Byoung-Lo Philo. *Two Koreas in Development*. New Brunswick, N.J.: Transaction, 1992.

Kim, Samuel S. "China and the Third World: In Search of a Peace and Development Line." In *China and the World*, edited by Samuel S. Kim, 192–204. Boulder, Colo.: Westview Press, 1989.

Kissinger, Henry A. *Diplomacy*. New York: Simon & Schuster, 1994.

———. *Nuclear Weapons and Foreign Policy*. New York: Harper & Brothers/Council on Foreign Relations, 1957b.

———. *White House Years*. Boston: Little, Brown, 1979.

———. *A World Restored: Metternich, Castlereagh and the Problems of Peace, 1812–1822*. New York: Houghton Mifflin, 1957.

———. *Years of Upheaval*. Boston: Little, Brown, 1982.

Klintworth, Gary. "Taiwan's International Identity: 400 Years in the Melting Pot." *Journal of East Asian Affairs (South Korea)* 10, no. 2 (1996): 373–94.

Kochavi, Noam. *A Conflict Perpetuated: China Policy During the Kennedy Years*. Westport, Conn.: Praeger, 2002.

———. "Limited Accommodation, Perpetuated Conflict: Kennedy, China, and the Laos Crisis, 1961–1963." *Diplomatic History* 26, no. 1 (2002): 95–135.

Koen, Ross. *The China Lobby in American Politics*. New York: Macmillan, 1960.

Koppes, Clayton R., and Gregory D. Black, *Hollywood Goes to War: How Politics, Profits and Propaganda Shaped World War II Movies*. Berkeley: University of California Press, 1987.

Lea, Homer. *The Valor of Ignorance*. New York: Harper, 1909.

League of Women Voters. *Taiwan and American Policy*. New York: Praeger, 1971.

Lee, Chin-Chuan. "Ambiguities and Contradictions: Issues in China's Changing Political Communication." In *China's Media, Media's China*, edited by Chin-Chuan Lee, 3–22. Boulder, Colo.: Westview Press, 1994.

Lee, Robert George. "Revolution and Foreign Relations in the Thought of Mao Zedong." Ph.D. diss., Brown University, 1980, 198.

Leviero, Anthony. "U.S. Not at War, President Asserts." *New York Times*, June 30, 1950, A1.

Li, Zhisui. *The Private Life of Chairman Mao.* Translated by Tai Hung-Chao. New York: Random House, 1994.

Lichterman, Martin. "Korea: Problems in Limited War." In *National Security in the Nuclear Age*, edited by Gordon B. Turner and Richard B. Challener, 31–56. New York: Praeger, 1960.

Lin, Cheng-Yi. "China, the U.S., and the Security of Taiwan." Ph.D. diss., University of Virginia, 1987.

Liu, Haiming. "Transnational Historiography: Chinese American Studies Reconsidered," *Journal of the History of Ideas* 65, no. 1 (2004): 135–53.

Liu, Xiaoyuan. *A Partnership for Disorder: China, the United States, and their Policies for the Postwar Disposition of the Japanese Empire, 1941–1945.* Cambridge: Cambridge University Press, 1996.

Liu, Yawei. "The United States according to Mao Zedong: Chinese-American Relations, 1893–1976." Ph.D. diss., Emory University, 1996.

Low, Alfred D. *The Sino-Soviet Confrontation Since Mao Zedong.* New York: Columbia University Press, 1981.

Lower, Peter. *Origins of the Korean War.* London: Longman, 1986.

Luce, Henry R. "The American Century," *Life*, February 17, 1941.

———. "Four Messages From China" (June 1941 speech for United China Relief). In *The Ideas of Henry Luce*, edited by John K. Jessup, 196–97. New York: Atheneum, 1969.

Lundberg, Ferdinand. *Imperial Hearst: A Social Biography.* Westport, Conn.: Greenwood, 1953.

Lyon, P. *Eisenhower: Portrait of the Hero.* Boston: Little, Brown, 1974.

Ma, Xiaohua. "The Sino-American Alliance During World War II and the Lifting of the Chinese Exclusion Acts." *American Studies International* 38 (2000): 39–61.

MacFarquhar, Roderick. *Sino-American Relations, 1949–1971.* New York: Praeger, 1972.

Madsen, Richard. *China and the American Dream: A Moral Inquiry.* Berkeley: University of California Press, 1995.

Mancall, M., ed. *Formosa Today.* New York: Praeger, 1964.

Mann, James. *About Face: A History of America's Curious Relationship with China, from Nixon to Clinton.* New York: Knopf, 1999.

Mao to Peng Dehuai. In "PLA attacks and Amphibious Operations," from *Jianguo Yilai Mao Zedong Wengao* [The manuscripts of Mao Zedong Since the founding of the nation. Translated by Shannon Wang. Beijing: Central Documents Publishing Co., 1987–90, 7:326.

Mao Zedong. *The Current Situation and the Task of the Party in 1949. Mao Zedong Junshi Wenxuan-Neibuban (Selected Works of Mao Zedong on Military Affairs, Edition for Internal Circulation).* Beijing: Jiefangjun Zhanshi Chubanshe, 1957.

———. *On the People's Democratic Dictatorship: Selected Works of Mao Tsetung.* Vol. 4. Beijing Foreign Languages Press, 1967.

———. "On the Ten Major Relationships," April 25, 1956. In *Selected Works of Mao Zedong,* Vol. 5. Beijing: Foreign Languages Press, 1977.

———. *Selected Works of Mao Tsetung.* Vol. 5. Beijing: Foreign Languages Press.

———. "The Strategies of CPVA," October 2, 1950. Translated by Suping Zhang. From *Collection of Mao Zedong Military Essays.* Beijing: Renmin Press, 1984, 6:106–8.

Martin, Bradley K. *Under the Loving Care of the Fatherly Leader: North Korea and the Kim Dynasty.* New York: St. Martin's Press, 2004.

Martin, Ralph G. *Henry and Clare: An Intimate Portrait of the Luces.* New York: Basic Books.

Mathews, Jay. "Edgar Snow Told You So; Why China's Great Leap Backward Should Come as No Surprise." *Washington Monthly* 21, no. 6 (July/August 1989): 50+.

Maud, Russell. "The Why of Nixon's Visits to China." *Far East Report,* July 1–8, 1972, 7.

Mayers, David A. *Cracking the Monolith: U.S. Policy Against the Sino-Soviet Alliance, 1949–1955.* Baton Rouge: Louisiana State University Press, 1986.

Mayers, David. *George Kennan and the Dilemmas of U.S. Foreign Policy.* New York: Oxford University Press, 1988.

McLean, David. "American Nationalism, the China Myth, and the Truman Doctrine: The Question of Accommodation With Peking, 1949–1950." *Diplomatic History* 10, no. 1 (1986): 30.

McNulty, Lawlor A. "The Sino-Soviet Border Problems of 1969." *Parameters* 13, no. 3 (1983): 59–67.

McSweeny, Dorothy Pierce. "F. Edward Hebert Oral History Interview I." Transcript, July 15, 1969, Lyndon Baines Johnson Library. 128.83.78.10/Johnson/Archives .Hom/Oralhistory.Hom/Hebert/Hebert.pdf.

Midgely, John J., Jr. *Deadly Illusions: Army Policy for the Nuclear Battlefield.* Boulder, Colo.: Westview Press, 1986.

Ming, Yuan. "The Failure of Perception: America's China Policy, 1949–1950." In *Sino-American Relations, 1945–1955: A Joint Reassessment of A Critical Decade,* edited by Harry Harding and Yuan Ming, 145–72. Wilmington, Del.: Scholarly Resources, 1989.

Murphy, John M. "Crafting the Kennedy Legacy." *Rhetoric & Public Affairs* 3, no. 4 (2002): 577–601.

Museum of the Chinese Revolution. *Album of Paintings.* Beijing: Cultural Relics Publishing House, 1991.

Nathan, Andrew J. *China's Crisis: Dilemmas of Reform and Prospects For Democracy.* New York: Columbia University Press, 1990.

———. "Re-Examining the Cold War: U.S.-China Diplomacy, 1954–1973." *China Review International* 10, no. 2 (2003): 436–40.

Neils, P. *China Images in the Life and Times of Henry Luce.* Savage, Md: Rowman & Littlefield, 1990.

Newhouse, John. *Cold Dawn.* New York: Holt, Rinehart and Winston, 1973.

Nixon, Richard M. "Asia After Vietnam." *Foreign Affairs* 46, no. 1 (1967): 111–25.

———. Excerpts from address to American Society of Newspaper Editors, April 20, 1963. In "Excerpts from Address by Nixon Calling for a Stronger U.S. Foreign Policy," *New York Times,* April 21, 1963, A62.

———. *RN: The Memoirs of Richard Nixon.* New York: Grosset and Dunlap, 1978.

———. *RN: The Memoirs of Richard Nixon.* Vol. 2. New York: Warner, 1979.

———. "The Speeches of Richard M. Nixon." Partial transcript of remarks, Knott's Berry Farm, Long Beach, Calif., October 12, 1960, 555–56.

———. "U.S. Foreign Policy For the 1970's: A New Strategy for Peace." A Report to the Congress, February 18, 1970, 3.

Nolt, James H. "Liberalizing Asia." *World Policy Journal* 16, no. 2 (Summer 1999): 94–110.

Norris, Robert S., William M. Arkin, and William Burr. "Where They Were." *Bulletin of the Atomic Scientists* 55, no. 6 (November/December 1999): 26–35. www.thebulletin.org/print.php?art_ofn-nd99norris_24.

NSC-6, "The Position of the U.S. Regarding Short-Term Assistance to China," March 26, 1948. Foreign Relations of the United States. 8: The Far East (China), 44–50.

NSC-68, "U.S. Objectives and Programs For National Security." In *Containment: Documents on American Policy and Strategy,* edited by Thomas H. Etzold and John Lewis Gaddis, 409. New York: Columbia University Press, 1978.

Ogata, Sadako. *Defiance in Manchuria: The Making of Japanese Foreign Policy, 1931–32.* Los Angeles: Center for Japanese and Korean Studies, 1964.

Ojha, I. C. "Sino-American Rapprochement: Why Wasn't It Predicted?" *China Report* 9, no. 3 (1973): 53–57.

Parker, Charles F. *Vietnam: Strategy for a Stalemate.* New York: Paragon Press, 1989.

Peck, Graham. *Two Kinds of Time.* Boston: Houghton Mifflin, 1950.

Peffer, Nathaniel. "Our Distorted View of China." *New York Times Magazine,* November 7, 1943.

Pei, Minxin. "Creeping Democratization in China." *Journal of Democracy* 6, no. 4 (1995): 65–79.

Perlmutter, David D. *Photojournalism and Foreign Policy: Framing Icons of Outrage in International Crises.* Westport, Conn.: Greenwood, 1998.

———. "Visual Images and Foreign Policy: Picturing China in the American Press, 1949–1989." Ph.D. diss., University of Minnesota, 1996.

Phillips, Cabell. *The Truman Presidency.* Baltimore: Penguin, 1969.

Phillips, Steve. "Confronting Colonization and National Identity: The Nationalists and Taiwan, 1941–45." *Journal of Colonialism and Colonial History* 2, no. 3 (2001).

Pollack, Jonathan. "The Korean War and Sino-American Relations." In *Sino-American Relations, 1945–1955: A Joint Reassessment of A Critical Decade,* edited by Harry Harding and Ming Yuan, 213–37. Wilmington, Del.: Scholarly Resources, 1989.

Powell, Lee Riley. *J. W. Fulbright and America's Lost Crusade.* Little Rock, Ark.: Roe Publishers, 1984.

Public Papers of the Presidents of the United States: Harry S. Truman, Containing the Public Messages, Speeches, and Statements of the President, 1945–1953. Truman address to U.S. Congress, September 1, 1950. Washington, D.C.: Government Printing Office, 609–14.

Public Papers of the Presidents of the United States: Richard Nixon 1972. Washington, D.C.: Government Printing Office, 1972.

Pye, Lucian. *China: An Introduction.* Boston: Little, Brown, 1978.

Qingdao Dashiji Shiliao, A Chronicle of Important Historical Events in Qingdao. Vol. 1. Qingdao: Qingdaoshi Danganguan, 1989.

Rahman, A. F. M. Shamsur. "United States Economic and Military Assistance Policy toward China During World War II and Its Immediate Aftermath." Ph.D. diss., University of Kansas, 1988).

Randle, Robert F. *Geneva 1954.* Princeton, N.J.: Princeton University Press, 1969.

Rees, David. *Korea: The Limited War.* New York: St. Martin's Press, 1964.

Rigg, Robert B. *China's Fighting Hordes.* Harrisburg, Pa.: Military Service Publishing Company, 1951.

Rising, George. *Clean for Gene: Eugene McCarthy's 1968 Presidential Campaign.* Westport, Conn.: Praeger, 1997.

ROC Yearbook 1955–56. Taipei: Government Information Office, 1956.

ROC Yearbook 1988. Taipei: Government Information Office, 1988.

Rogers, F. E. "Sino-American Relations and the Vietnam War, 1964–1966." *China Quarterly* 66 (1976): 293–314.

Romanus, Charles F., and Riley Sunderland. *Time Runs Out in CBI.* Washington, D.C.: Office of the Chief of Military History, Department of the Army, 2003.

Romanus, Charles F., and Riley Sunderland. *Stilwell's Mission to China.* Washington, D.C.: Department of the Army, 1953.

———. *United States Army in World War II: China-Burma-India Theater: Stilwell's Command Problems.* Washington, D.C.: Office of the Chief of Military History, Department of the Army, 1956.

Rosenberg, Jonathan. "Between 'Paper' and 'Real Tigers': Mao's View of Nuclear Weapons." In *Cold War Statesmen Confront the Bomb: Nuclear Diplomacy Since 1945,* edited by John Lewis Gaddis, Philip H. Gordon, Ernest R. May, and Jonathan Rosenberg, 171–93. Oxford: Oxford University Press, 1999.

Rosendorf, Neal. "John Foster Dulles's Nuclear Schizophrenia." In *Cold War Statesmen Confront the Bomb: Nuclear Diplomacy Since 1945,* edited by John Lewis Gaddis, Philip H. Gordon, Ernest R. May, and Jonathan Rosenberg, 62–86. Oxford: Oxford University Press, 1999.

Rostow, Walt W. "Robert S. Mcnamara Oral History Interview I." Transcript, January 8, 1975, Lyndon Baines Johnson Library. 128.83.78.10/johnson/archives.hom/oral history.hom/mcnamarar/mcnamara1.pdf.

Schaller, Michael. "Consul General O. Edmund Clubb, John P. Davies, and the Inevitability of Conflict Between the United States and China, 1949–1950: A Comment and New Documentation." *Diplomatic History* 9, no. 2 (1985): 149–53.

———. *U.S. Crusade in China: 1938–1945.* New York: Columbia University Press, 1979.

Schandler, Herbert Y. *The Unmaking of a President: Lyndon Johnson and Vietnam.* Princeton, N.J.: Princeton University Press, 1977.

Schlesinger, Arthur A. *A Thousand Days: John F. Kennedy in the White House.* Boston: Houghton Mifflin, 1965.

Service, John S. *The Amerasia Papers: Some Problems in the History of U.S.-China Relations.* Berkeley: University of California, 1971.

Seymour, James D. "Sanctions or Subdued Relations: The International Response to the 1989 Massacre." *Fletcher Forum of World Affairs* 14, no. 1 (1990): 55–61.

Shanghai Communiqué, February 27, 1972. 1974. *Public Papers of the Presidents of the United States: Richard Nixon, 1972.* Washington, D.C.: Government Printing Office.

Sheng, Michael W. *Battling Western Imperialism: Mao, Stalin, and the United States.* Princeton, N.J.: Princeton University Press, 1997.

Shewmaker, Kenneth E. *Americans and Chinese Communists, 1927–1945.* Ithaca, N.Y.: Cornell University Press, 1971.

Shirk, Susan. "1977–1978. Human Rights: What about China?" *Foreign Policy,* no. 29 (Winter 1987): 109–27.

Simon, Nancy S. "From the Chinese Civil War to the Shanghai Communiqué: Changing U.S. Perceptions of China as a Security Threat." Ph.D. diss., Johns Hopkins University, 1982.)

Small, M. "Public Opinion." In *Encyclopedia of American Foreign Policy: Studies of the Principal Movements and Ideas,* edited by A. Deconde, 844–55. New York: Scribner, 1978.

Snow, Edgar. "China Will Talk from a Position of Strength." *Life,* July 30, 1971, 24.

Sorensen, Theodore C. *Kennedy.* New York: Harper & Row, 1965.

Spanier, John. *American Foreign Policy Since World War II,* 11th rev. Washington, D.C.: Congressional Quarterly, 1988.

"Special Counsel to the President Charles Murphy." Oral History Interview, May 21, 1969. www.trumanlibrary.org/wake/meeting.htm.

Spence, Jonathan. *To Change China: Western Advisers in China, 1620–1960.* Boston: Little, Brown, 1969.

Srinivas Chary, M. *The Eagle and the Peacock: U.S. Foreign Policy toward India Since Independence.* Westport, Conn.: Greenwood, 1995.

Stevenson, Adlai. "Putting First Things First: A Democratic View." *Foreign Affairs* 38, no. 3 (1960): 191–208.

Stilwell, Joseph W. *The Stilwell Papers.* Arranged and edited by Theodore H. White. New York: William Sloane Associates, Inc., 1948.

Stolper, Thomas E. *China, Taiwan, and the Offshore Islands.* Armonk, N.Y.: M. E. Sharpe, 1985.

Stone, I. F. *The Hidden History of the Korean War.* New York: Monthly Review Press, 1952.

Strong, Tracy B., and Helene Keyssar. *Right in Her Soul: The Life of Anne Louise Strong.* New York: Random House, 1983.

Stueck, William W., Jr. "American Policy toward China and Korea, 1947–1950" Ph.D. diss., Brown University, 1977.

———. *The Road to Confrontation: American Policy toward China and Korea, 1947–1950.* Chapel Hill: University of North Carolina Press, 1981.

———, ed. *The Korean War in World History.* Lexington: University Press of Kentucky, 2004.

Summers, Harry G., Jr. *On Strategy: A Critical Analysis of the Vietnam War.* San Francisco: Presidio Press, 1982.

Sutter, Robert. *China Watch: Toward Sino-American Reconciliation.* Baltimore: Johns Hopkins University Press, 1978.

Swanberg, W. A. *Luce and His Empire.* New York: Charles Scribner's Sons, 1972.

Szulc, Tad. "Zhou Said to Back Parley with U.S." *New York Times,* June 18, 1970, A9.

Tang, Tsou. *America's Failure in China, 1941–50.* Chicago: University of Chicago Press, 1963.

Tebbel, John. *The Life and Good Times of William Randolph Hearst.* New York: Dutton, 1970.

Thomson, James, Jr. "On the Making of U.S. China Policy, 1961–1969: A Study in Bureaucratic Politics." *China Quarterly*, no. 50 (April–June 1972): 222–23.

Thorne, Christopher. *Allies of a Kind: The United States, Britain, and the War against Japan.* Oxford: Oxford University Press, 1978.

Tocquevill, Alexis de. Democracy in America. Edited by J. P. Mayer. Translated by G. Lawrence. Garden City, N.Y.: Doubleday, 1969.

Truman, Harry S. *Memoirs.* New York: Doubleday, 1955[1956].

Tsai, George Woei. "The Taiwan Straits Crises: Analysis of the China, Taiwan, United States Relationship, 1949–1983." Ph.D. diss., Northern Arizona University, 1985.

Tsou, T. "Statesmanship and Scholarship." *World Politics* 26, no. 3 (1974): 428–51.

Tu, W. "Cultural China: The Periphery as the Center." *Dædalus* 120 (1991): 1–32.

Tuchman, Barbara W. *Stillwell and the American Experience in China, 1911–45.* New York: Bantam, 1972.

Tucker, Nancy B. "China as a Factor in the Collapse of the Soviet Empire." *Political Science Quarterly* 110 (1995): 501–18.

———. *Patterns in the Dust: Chinese-American Relations and Recognition Controversy, 1949–1950.* New York: Columbia University Press, 1983.

———, ed. *China Confidential: American Diplomats and Sino-American Relations 1945–1996.* New York: Columbia University Press, 2001.

U.S. News & World Report. September 16, 1968, 48.

Ulam, Adam. *The Rivals: America and Russia Since World War II.* New York: Penguin, 1980.

United States Department of State. *Foreign Relations of the United States: Diplomatic Papers, 1949–1973,* Washington, D.C.: Government Printing Office.

Vandenberg, Arthur, Jr., and Arthur H. Sense. *The Private Papers of Senator Vandenberg, 1944–1951.* Boston: Houghton Mifflin, 1952.

"Vice President Richard M. Nixon and Senator John F. Kennedy, Second Joint Radio-Television Broadcast." October 7, 1960, originating NBC, Washington, D.C. www.jfklibrary.org/60-2nd.htm.

Volume of Files on Sino-U.S. Relations. Vol. 2, Translated by Suping Zhang. Beijing: Renmin Press, 1984, 91–92.

Wang, Chi-Wu. "Military Preparedness and Security Needs: Perceptions from the Republic of China on Taiwan." *Asian Survey* 21, no. 6 (1981): 651.

Wang, De. "The Battle to Liberate the Dachen Isles." In *San Jun Huizhan (The Land, Air and Naval Fighting in the East China Sea),* edited by Fengzhi Nie, 2. Nanjing: Jiefangjun Chubanshe, n.d.

Wang, Yufeng. "John S. Service and United States-China Relations, 1933–1945." Ph.D. diss., University of West Virginia, 1997.

Weigley, R. F. *The American Way of War: A History of United States Military Strategy and Policy.* New York: Macmillan, 1973.

Wheelwright, Philip. *The Burning Fountain: A Study in the Language of Symbolism.* Bloomington: Indiana University Press, 1968.

White, Theodore H. *In Search of History: A Personal Adventure.* New York: Harper & Row, 1978.

Whiting, Allen S. *China Crosses the Yalu.* New York: Macmillan, 1960.

———. "China's Use of Force, 1950–96, and Taiwan." *International Security* 26, no. 2 (2001): 103–31.

———. "Mao, China, and the Cold War." In *The Origins of the Cold War in Asia,* edited by Akira Iriye and Yonosuke Nagai, 249–81. New York: Columbia University Press, 1977.

———. "Quemoy 1958: Mao's Miscalculations." *The China Quarterly* 62 (June 1975): 263–70.

Wilson, David L. "The Attitudes of American Consular and Foreign Service Officers toward Bolshevism in China, 1920–1927." Ph.D. diss., University of Tennessee, 1974.

Wu, Lengxi. "Inside Story of the Decision Making During the Shelling of Jinmen." Translated by Shannon Wang. *Zhuanji Wenxue* [Biographical literature], no.1 (1994): 11.

Xia, Yafeng. "From Antagonists to Adversaries: United States-China Talks During the Cold War, 1949–1972." Ph.D. diss., University of Maryland, 2003.

Xiong, Xianghui. *Wo De Qingbao Yu Waijiao Shengya* [My intelligence and diplomatic career]. Beijing: Zhongyang Dangxiao Chubanshe, n.d., 184–86.

Xu, Guangqiu. "The Issue of U.S. Air Support for China During the Second World War, 1942–45." *Journal of Contemporary History* (Great Britain) 36, no. 3, (2001): 459–84.

Yeh, Chiou-Ling. "In the Traditions of China and in the Freedom of America: The Making of San Francisco's Chinese New Year Festivals." *American Quarterly* 56, no. 2 (2004): 395–420.

Yibo, Bo. *Some Decision Making in Retrospect.* Vol. 1. Translated by Suping Zhang. Zhong Gong Zhong Yang Dang Xiao Press. Beijing: Renmin Press.

Yim, Kwan Ha, ed. *China and the U.S. 1955–63.* New York: Facts on File, 1973.

Yoshihashi, T. *Conspiracy at Mukden: The Rise of the Japanese Military.* New Haven, Conn.: Yale University Press, 1963.

Young, Kenneth T. *Negotiating with the Chinese Communists: The U.S. Experience, 1953–67.* New York: McGraw-Hill, 1968.

Zagoria, Donald S. *The Sino-Soviet Conflict, 1956–1961.* Princeton, N.J.: Princeton University Press, 1962.

Zhai, Qiang. *China and the Vietnam Wars, 1950–1975.* Chapel Hill: University of North Carolina Press, 2000.

———. "Cold War International History Project, Virtual Archive. Beijing and the Vietnam Conflict, 1964–1965: New Chinese Evidence." Translated by Qiang Zhai. wwics.si.edu/index.cfm?topic_id=1409&fuseaction=library.document&id=244.

———. "Recent Chinese Writings on 1945–1955 Sino-U.S. Relations." *SHAFR Newsletter* 20(4) (1989): 75–84.

Zhang, Baijia. "Kang Ri Zhanzheng Shiqi Guogong Liangdang De Dui Mei Zhengce." [Guomindang and Communist policies toward the United States during the anti-Japanese war]. In Zhong Mei Guanxi Shi Shang Chenzhong De Yiye [Sino-American relations, 1944-1955: Collaborative reassessment of a troubled time], edited by Yuan Ming and Harry Harding, 19–20. Beijing: Beijing University Press, n.d.

Zhang, Xiaoming. "China and the Korean War: From a Chinese Perspective." *Journal of Asiatic Studies* [Asea Yon'gu] (South Korea) 41, no. 1 (1988): 63–80.

Zheng, Guoqiang. "Conflict, Contact, and Coexistence: Sino-American Relations, 1948–1960." Ph.D. diss., University of Toledo, 1997.

Zhonghua Renmin Gongheguo Duiwai Guanxi Wenjianji, 1949–1950. [Documents of foreign relations of the People's Republic of China, 1949–1950]. Translated by Shannon Wang. Beijing: World Knowledge Publishing House, 1957, 130.

Zhou, Enlai. "Assisting Korea to Fight U.S. Is Still the Primitive Mission in 1951." April 2, 1951. In *Selection of Zhou Enlai's Military Essays.* Beijing: Renmin Press, 1997.

———. "Be Prepared to Win." In *Selection of Zhou Enlai's Military Essays,* translated by Suping Zhang, 2:43–44. Beijing: Renmin Press, 1984.

———. *Report On Problems Concerning Peace Talks. Selected Works of Zhou Enlai, Volume L* (Beijing: Foreign Language Press, 1981), 360–64.

———. "Struggle To Solidify and Develop People's Victory." September 30, 1950. In *Selection of Zhou Enlai's Military Essays,* translated by Suping Zhang, 2:37. Beijing: Renmin Press, 1984, 37.

Zweiback, Adam J. "The 21 'Turncoat GIs': Nonrepatriations and the Political Culture of the Korean War." *Historian* 60, no. 2 (1998): 345–62.

Index

About the Author

David D. Perlmutter is a professor and associate dean for graduate studies and research in the William Allen White School of Journalism and Mass Communications at the University of Kansas. A documentary photographer, he is the author of three books and the editor of a fourth on war, politics, visual images, new media, and public opinion: *Photojournalism and Foreign Policy: Framing Icons of Outrage in International Crises* (1998); *Visions of War: Picturing Warfare from the Stone Age to the Cyberage* (1999); (ed.) *Guide to Political Communication* (1999); and *Policing the Media: Street Cops and Public Perceptions of Law Enforcement* (2000). A former board member of the American Association of Political Consultants, he is a co-founder of the Association of Schools of Political Campaign Management. He writes frequently for the *Chronicle of Higher Education.* He has written over 130 opinion essays for U.S. and international newspapers. In addition, he is frequently interviewed by wire services, newspapers, magazines, and television media such as the *New York Times,* National Public Radio, CNN, Reuters, UPI, the AP, and ABC News, as well as having been featured in a number of documentaries.

www.ingramcontent.com/pod-product-compliance
Lightning Source LLC
LaVergne TN
LVHW011749200926
848484LV00003B/662
9780739118191